Lean Sustainability

Creating Safe, Enduring, and Profitable Operations

Lean Sustainability

Creating Safe, Enduring, and Profitable Operations

Dennis Averill

CRC Press
Taylor & Francis Group
Boca Raton   London   New York

CRC Press is an imprint of the
Taylor & Francis Group, an **informa** business
A PRODUCTIVITY PRESS BOOK

CRC Press
Taylor & Francis Group
6000 Broken Sound Parkway NW, Suite 300
Boca Raton, FL 33487-2742

© 2011 by Taylor & Francis Group, LLC
CRC Press is an imprint of Taylor & Francis Group, an Informa business

No claim to original U.S. Government works

Version Date: 20110708

International Standard Book Number: 978-1-4398-5716-8 (Paperback)

Library of Congress Cataloging-in-Publication Data

Averill, Dennis.
    Lean sustainability : creating safe, enduring, and profitable operations / Dennis Averill.
        p. cm.
    Includes bibliographical references and index.
    ISBN 978-1-4398-5716-8 (pbk. : alk. paper)
    1. Lean manufacturing. 2. Industrial safety. 3. Industrial hygiene. 4. Industrial management--Environmental aspects. I. Title.

TS155.A87 2011
670--dc23   2011026927

**Visit the Taylor & Francis Web site at
http://www.taylorandfrancis.com**

**and the CRC Press Web site at
http://www.crcpress.com**

To the giant, Ed Averill; it was only by
standing on his shoulders that I was able
to see farther and more clearly.

Contents

Preface

For many years it was a common belief among American business leaders that it cost extra to make a quality product. For some companies, most notably those in the American automobile industry, this mistaken belief led to shrinking sales, lost market share, vanishing profits, and in a few cases almost to the ultimate demise of the entire business enterprise. After suffering a few hard lessons at the hands of global competitors, it is commonly accepted among today's titans of industry that outstanding product quality is not a luxury but a necessary and integral ingredient for sustained business success. To ignore product quality is considered folly, and a quick path to the unemployment line.

Unfortunately, some business leaders still contend that protecting employees and the environment, and conducting their operations in a sustainable fashion are barriers to business success. These myopic managers maintain that safety and sustainability are extra work that require additional resources and divert business efforts away from the primary job of building efficient and profitable operations. The purpose of this book is to dispel the myth that safety and sustainability are contrary to business success. The hope is that after reading, reflecting upon, and applying the principles and approaches advocated in this volume it will become eminently clear that any business person who ignores safety and sustainability is as imprudent as one who ignores quality.

The keys to achieving safe, sustainable, and profitable operations are integrating and leveraging Lean methodologies in all areas of the business. Safety and sustainability are not additional or separate work, but rather, they are the way one runs a "Lean, Green, and Serene" enterprise. Lean, SHE (safety, health, and environmental protection), and sustainability focus on similar objectives:

1. Eliminating accidents, incidents, waste, and losses
2. Increasing operational efficiency
3. Conducting business in a sustainable way that conserves resources and reduces the business's environmental footprint

By linking an organization's Lean, SHE, and sustainability processes, natural synergy and efficiency are created that benefit all areas of the business and offer the enterprise a real prospect of achieving sustained profitable growth, or, as is commonly expressed, "The opportunity to do well by doing good." However, as with any complex business endeavor, realizing the vision of "Lean, Green, and Serene" is easier said than done. In other words, the devil is in the details. Excellence in productivity, SHE, and sustainability are not achieved by exhortations to do better, or by setting meaningless targets, but rather by doing the hard work to implement business systems that systematically and relentlessly identify and eliminate risk, waste, and losses from the workplace.

Many books and periodicals have been written on the theory of Lean and the conceptual basis of SHE excellence and sustainability. However, few resources are available that provide practical, detailed, real-world methodologies and tools for integrating these three disciplines and realizing safe, sustainable, and profitable operations. My hope is that this book will be the key for your organization that unlocks the vault containing the secret recipe for safe, sustainable business success. This publication is the product of over 20 years of experience with implementing Lean, SHE, and sustainability processes in the chemical and consumer products industries. I hope you find the hints and lessons I share to be helpful for you and your organization's journey toward "triple zero": zero accidents, zero incidents, and zero losses.

It is my firm belief that any organization that hopes to call itself great must value people, the planet, and posterity. The "Lean, Green, and Serene" philosophy and approach detailed herein provide a path to organizational greatness, where a business can do well financially by doing good for people and the planet. I believe if you follow this less-traveled path you and your organization will achieve sustained profitable growth. You, your family, your fellow employees, the planet, and future generations will be glad that you took up the challenge to be "Lean, Green, and Serene."

Acknowledgments

Virtually everything I have learned I have learned from others. This is certainly true of my knowledge related to Lean, SHE, and sustainability. My first safety teachers were my parents and my siblings, to whom I am most indebted. Without their frequent and timely intervention and sincere safety admonitions in my formative years, my passion for safety and sustainability may never have blossomed. This is doubly true for my father, Ed Averill, who not only provided a healthy dose of fatherly advice, but as an industrial hygienist and SHE professional for Mobil Oil provided me with an interest in manufacturing and a passion for safety at a young age. As a novice professional in the field following in his footsteps, he served as a much-valued professional mentor.

I also learned much from many of the SHE and Lean professionals I have trained and worked with over the years who were unselfish in their advice and their willingness to engage me in lively, and I'm sure at times tiresome, debate on technical issues. Although they are too numerous to mention, I will take the time to credit those who have had a significant impact on my professional development, and on the crystallization of my thoughts related to Lean, SHE, and sustainability. I am indebted to my graduate school instructors and colleagues, Dr. Morton Corn, Dr. Charles Billings, and Dr. Patrick Breysse of the Johns Hopkins Bloomberg School of Public Health. I am grateful to my professional colleagues and confidants, Ned Berg (classmate and consultant), Bernie Silverstein (Brookhaven National Laboratory), Chuck Brehm (American Cyanamid), and Alan Weikert (American Cyanamid and W. L. Gore), who have always been available for advice and counsel. I am most thankful to past colleagues at Unilever, Lou Piombino, Maria Ruibal, and Dirk Lueders, who over the years served as a sounding board for my ideas. I am particular indebted to Randy Mosebrook of Sun Products Company, who provided helpful advice and support on this project. In the area of Lean and TPM I am most grateful to Shinichi Shinotsuka, who served as my TPM/Lean instructor and sensei, and also to my fellow Unilever TPM instructors, Edna DeFlavis, Cindy Rigby, and Mike Johnston, who spent long weeks away from home leading instructors' courses and living and breathing the Lean life. To paraphrase Bernard of Chartres and Sir Isaac Newton, if you find my work enlightening it is only because "like a dwarf it is only by standing on the shoulders of giants that I have been able to see farther."

Thanks to my publisher, Taylor & Francis Group/Productivity Press, for its belief in this project and willingness to back a young author. Finally, a hug and heartfelt thanks to my wife, Joan, and my children, Brian and Caitlin, for their support and understanding while I endeavored to write this book. Many evenings and weekends they were left to fend on their own while I hunkered down in the college library. Their words of encouragement were the fuel that kept me going and the laptop humming.

About the Author

Dennis Averill CIH, CSP has over 25 years of management experience in the chemical, food, and consumer products industries, leading corporate programs in the areas of safety, health & environment (SHE), quality, and manufacturing improvement (Lean & TPM).

Mr. Averill was a Phi Beta Kappa, Bachelor of Science graduate of the University of Richmond.

He has also earned a Master of Health Science degree in environmental health engineering, and a Master of Administrative Science degree in business management from the Johns Hopkins University where he has served as an associate of the Johns Hopkins Bloomberg School of Public Health. He is a Certified Industrial Hygienist (CIH), a Certified Safety Professional (CSP), and a Certified TPM Instructor.

He has lent his time and talents to various professional and community groups serving as president of the American Industrial Hygiene Association, Chesapeake Section; vice president of the Community Coalition of Harford County; member of the Harford County Emergency Planning Committee; industry representative on the Maryland Governor's Council on Toxic Substances; and member of the American Society of Safety Engineers Management Specialty Practice Group.

Chapter 1

Safety, Health, and Environmental (SHE) Pillar: Foundation of Lean Continuous Improvement

Safe upon solid rock the ugly houses stand.

Edna St. Vincent Millay
American poet 1842–1950

1.1 Beginnings of Lean and TPM

Lean production is an improvement model and collection of tools that emphasizes the elimination of all types of waste (*muda*) and non-value-added activities, and the delivery of high-quality products at the lowest possible cost. The focus of Lean is on producing more goods with fewer resources by driving continual improvement in all areas of business performance, including cost, productivity, efficiency, and safety, health, and environment (SHE). Key principles of a Lean supply chain include

1. Value is defined by the customer.
2. The supply chain and value stream should flow continuously.
3. The entire organization must manage toward perfection by eliminating waste and adding value.[1,2]

It is important to note that from a Lean perspective, workplace accidents and environmental incidents are wastes that harm the value stream, hamper continuous flow, and work against the goal of perfection. Therefore, when implemented properly, Lean production is consistent with, and supportive of, SHE excellence.

To fully appreciate the impact of Lean thinking and practice on modern manufacturing and its relationship to achieving SHE excellence, it is helpful to have an understanding of its history, roots, and philosophy. Although the modern Lean methodology is derived from the Toyota Production System (TPS), its history and evolution can be traced back almost 100 years. American industrialist Henry Ford is commonly recognized as the first person to move away from craft production and use the assembly line to achieve production flow and mass production. In 1913 at his Model T automobile manufacturing complex in Highland Park, Michigan, Ford integrated the concepts of interchangeable parts, standard work, and production flow via the moving assembly line. Although Ford's mass production process successfully produced over 15 million Model T autos at low cost, his system had some marked deficiencies. Specifically, Ford's production process was inflexible and unable to provide product variety. As Ford wrote in his autobiography, "Any customer can have a car painted any colour that he wants so long as it is black."[3] Ford's mass production approach "only worked when production volumes were high enough to justify high-speed assembly lines, when every product used exactly the same parts, and when the same model was produced for many years."[4] Although Ford was a pioneer in advancing better pay for workers, Ford's system also suffered from the fact that it tended to devalue workers, assigning them to repetitive tasks, not tapping their full capabilities, and not seeking their input. To a large extent, labor was viewed solely as a cost of production, and not something that could add value. Workers on the production line were replaceable and expendable. Although the mass assembly line resulted in marked production improvements and some improvements in safety, working conditions were far from ideal.[5]

The Ford Motor Company's mass production model played a pivotal role in the Allied victory during World War II. The ability of American industry to produce the massive quantity of war materiel that ensured victory did not go unnoticed by Japanese industrialists. At the Toyota Company, Eiji Toyoda, Shigeo Shingo, and Taiichi Ohno studied Ford's production system, American grocery chains, and the statistical process control methods of W. Edwards Deming, Kaoru Ishikawa, and Joseph Juran. Around 1950, based upon these studies of American production methods, Toyota's chief engineer, Taiichi

Ohno, developed the Toyota Production System. The main objectives of the Toyota system are to eliminate waste (*muda*), stress (*muri*), and inconsistency (*mura*) by creating a smooth, flexible, accident-free, and efficient production system. Through key approaches such as just-in-time (JIT), error-proofing (*poka-yoke*), autonomation (*jodoka*), standardized work, and continuous improvement (*kaizen*); waste, stress, accidents, and inconsistency are eliminated and efficient, safe, high-quality, low-cost production is achieved. The Toyota Production System and modern Lean production both value factory workers for their muscle and their brains, and tap into employee creativity and ingenuity by involving everyone in continual improvement activities. According to Toyota, to function well, a factory must optimize both its equipment and its people. "The power behind TPS is a company's management commitment to continuously invest in its people and promote a culture of continuous improvement."[6] Toyota's approach to manufacturing, termed the Toyota Way, is summarized in its four high-level principles:

1. Go and See for Yourself (*genchi genbutsu*)
2. Continuous Improvement (*kaizen*)
3. Respect and Teamwork
4. Challenge

Consistent with the principle of respect and teamwork, the Toyota Production System emphasizes workplace safety, health, and environment. Organizations using TPS typically measure their progress by using the acronym QCDSM to focus on key performance indicators in the areas of quality, cost, delivery, safety, and morale. SHE excellence is consistent with the TPS goal of creating an efficient, stress-free work environment.[7] In an internal Toyota communication, Taiichi Ohno expressed the organization's uncompromising commitment to safety excellence: "Every method available for man-hour reduction to reduce cost must, of course, be pursued vigorously; but we must never forget that safety is the foundation of all our activities. There are times when improvement activities do not proceed in the name of safety. In such instances, return to the starting point and take another look at the purpose of that operation."[8]

1.2 James Womack and Lean

James Womack and fellow MIT researcher John Krafcif are credited with coining the term "Lean production." During the mid- and late 1980s, James Womack and other researchers at MIT's International Motor Vehicle Program studied the manufacturing systems of global auto manufacturers. They concluded that the auto industries of North America continued to rely upon outdated mass production systems, whereas Japanese companies were employing a new technique that they termed "Lean production" because it used significantly fewer resources than mass production. According to Womack, "Lean production combines the advantages of craft and mass production.... Lean producers employ teams of multiskilled workers at all levels of the organization and use highly flexible, increasingly automated machines to produce volumes of products in enormous variety."[9] Lean organizations are focused on perfection and committed to continuous improvement. The goals are continually declining costs, zero defects, zero losses, and zero accidents. Womack argued that a fundamental shift in global manufacturing methods had taken place, characterized by the decline of mass production and the dawning of a new age of Lean production. Unlike mass production, Lean production valued workers for more than their brawn. Workers were grouped in teams and were expected to take ownership of equipment and to assume responsibility for all operations in their work area including housekeeping, minor repairs, quality production, safety, and process improvement.[10]

1.3 TPM

In the 1970s a management improvement methodology known as total productive maintenance was developed by the Japanese based upon American preventive maintenance or PM concepts. In 1971 the Nippondenso Co., Ltd., a manufacturer of automobile parts for Toyota, received the PM award for its total-member-participation PM program or TPM process. With the assistance of the Japan Institute of Plant Maintenance (JIPM), the Nippondenso/Toyota process gradually evolved into the current total productive maintenance or TPM methodology. The following key characteristics of the TPM methodology reveal its close relationship to Lean production:

1. The pursuit of overall organizational efficiency via the persistent elimination of all losses
2. Ownership of equipment, processes, and associated losses by operators
3. Implementation of continuous improvement via overlapping small group activities
4. A hands-on approach to build a zero accident, zero defect, zero loss system throughout the organization

Originally the Japanese TPM methodology consisted of five areas of focus, or pillars: focused improvement (FI or *Kobetsu Kaizen*), autonomous maintenance (AM or *Jishu Hozen*), planned maintenance (PM or *Keikaku Hozen*), training and education (TE), and early management (EM). TPM's initial focus on production activities and associated losses, although beneficial, was incomplete. It became clear that a process that did not consider quality, management, or safety, health, and environmental losses could not sustain true continuous improvement. Therefore, as indicated in Figure 1.1, three additional pillars were added to the TPM process: quality maintenance (QM or *Hinshitsu Hozen*), TPM in administration (the office), and safety, health, and environment (SHE). Today the eight pillars of TPM are commonly applied companywide in both production and nonproduction areas such as engineering, administration, development, and distribution. In light of this, some organizations have indicated the expanded scope of their continuous improvement process by changing the meaning of the TPM acronym from "Total Productive Maintenance" to the more expansive "Total Productive Manufacturing" or "Total Perfect Manufacturing."[11]

Today, the Japan Institute of Plant Maintenance (JIPM) requires organizations to reduce accidents and environmental pollution in order to receive its TPM Excellence Award, also known as the PM Prize. Safety, health, and

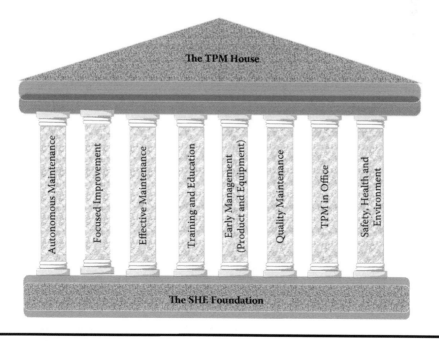

Figure 1.1 The pillars of TPM.

environmental management is an essential component of an effective TPM process. TPM supports the goal of SHE excellence in the following ways:

1. It inculcates the vision and philosophy of zero accidents and zero environmental incidents into people's way of thinking.
2. TPM integrates the goal of zero accidents and zero environmental incidents into the organization's way of working.
3. Through small group activities TPM involves employees in making their equipment and work areas safer.
4. Through autonomous maintenance activities workers take ownership of their work area and become knowledgeable about all aspects of their equipment, including safety.
5. Through the autonomous maintenance, planned maintenance, and focused improvement processes equipment is made more reliable and safer.
6. By employing focused improvement methodologies SHE improvement projects (*kaizens*) are implemented and an attitude of continual SHE improvement is fostered.

1.4 The Safety, Health, and Environmental (SHE) Pillar

The SHE pillar is the gateway to all continuous improvement, regardless of which specific process an organization uses, whether it is Lean, TPM, or another improvement methodology. If employees do not believe that an organization is doing right by them in the area of safety, health, and environment it is unlikely that they will cooperate with or actively participate in any continuous improvement process. According to Abraham Maslow's concept of the hierarchy of needs outlined in his paper "A Theory of Human Motivation,"[12] people cannot address higher-order needs if their basic needs are not satisfied. As shown in Figure 1.2, Maslow identified safety as a basic human need that must be met first. Therefore, if employees, as a result of accidents and uncontrolled risks, feel that they are unsafe in the workplace they will tend to devote their energies to meeting their basic human need for safety rather than undertake more advanced improvement activities that involve creativity and problem-solving. In a similar fashion, the Japan Institute of Plant Maintenance defines safety as "the maintenance of peace of mind."[13] Without the peace of mind or serenity brought about by a safe working environment, employees are unwilling and even unable to focus their energies on production improvement activities. Thus, it can be said that all improvement begins with safety.

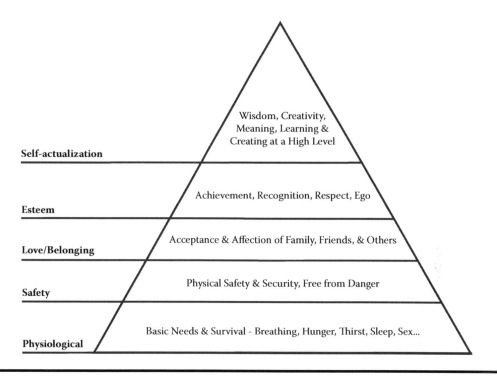

Figure 1.2   Maslow's hierarchy of needs.

The safety, health, and environmental pillar is the cornerstone of the Lean or total perfect manufacturing process because it lends support and stability to the overall continual improvement process and to all of the other pillars. Safety, health, and environment is a natural area in which to begin a continuous improvement initiative inasmuch as all employees can accept that they will benefit from SHE improvement. Typically, employees will buy into and actively participate in SHE improvement activities. The organizational discipline, attention to detail, and teamwork needed for a successful SHE pillar will provide a strong foundation for the other continuous improvement pillars. A sturdy, well-developed SHE pillar is essential to realizing the Lean and TPM goals of:

- Achieving zero accidents, zero incidents, and zero losses
- Building a world-class supply chain
- Creating a lean, green, and serene organization

It is obvious that a manufacturing or supply chain improvement process that results in the injury of personnel or damage to the environment is fundamentally flawed and will not deliver meaningful long-term improvement.

Indeed, production improvements that harm people or the environment are not really improvements at all, but simply a method of substituting one loss for another loss. From a Lean perspective, accidents and environmental incidents are waste that do not add value and hamper the organization's journey toward perfection.

As depicted in Figure 1.3, the Lean/TPM process can be looked at as the engine of continual improvement, and SHE can be considered the essential lubricant that enables the engine to function properly. The various pillars of the Lean/TPM methodology are the key components of the continual improvement engine that must work together like a set of finely meshed gears if they are to work effectively and efficiently. As with any engine, the continuous improvement engine will not function properly and will ultimately fail without an adequate supply of lubricant provided by the safety, health, and environmental process. The activities of the SHE pillar, like those of the other TPM pillars, must be driven by loss data, the main drive gear. Few organizations have unlimited resources; therefore, the SHE pillar should focus on activities that reduce losses: accidents and environmental incidents.

Safety, health, and environmental improvement is a good place to begin an overall continuous process because the goals of the Lean/TPM process and SHE are the same. The key goal of Lean/TPM, and the SHE pillar in particular, is the elimination of all losses. If world-class performance is to be achieved in Lean or SHE, the organization must have an unrelenting passion for the identification and elimination of all losses. SHE excellence requires discipline, attention to detail, and teamwork—organizational characteristics that are also essential for Lean success. Each site safety, health, and

Figure 1.3   The Lean TPM machine.

environmental pillar team should ask itself if it is truly on the path to SHE excellence as indicated by the organization's passion for eliminating safety, health, and environmental losses. Each SHE pillar team should ask whether the site or organization has a

- Comprehensive SHE loss tree:
  - Does the site regularly publish a SHE loss tree (inventory of organization's losses) that details all the site's key safety, health, and environmental losses? Are the losses quantified by both number and dollar amount?
- Passion and sense of urgency about eliminating SHE losses:
  - Does everyone at the site know the site SHE losses, and are they actively working to eliminate them?
- Formal SHE *kaizen* process:
  - Does the organization have a formal process for establishing SHE improvement (*kaizen*) teams that are chartered to eliminate specific SHE losses?
  - How many SHE projects are in the current site focused improvement funnel?
  - How many SHE *kaizen* or improvement teams have been established this year?
- SHE pillar plan focused on loss elimination:
  - Are all of the site's SHE activities (training, inspections, OPLs, risk assessments, etc.) conducted with the organization's loss tree in mind?
  - Are the organization's limited SHE resources focused on eliminating its major safety, health, and environmental losses, hazards, and risks?
- A process to learn about all SHE losses:
  - Is a thorough root cause analysis conducted for all accidents and incidents?
  - Is there a burning desire to learn from all site incidents and to prevent their recurrence?
  - Does the organization conduct 30–60–90 day checks to ensure that the follow-up SHE actions have been effective?

Initiating a continuous improvement process with the SHE pillar also makes sense because many of the tools and techniques used in Lean/ TPM are quickly applied to safety, health, and environment because similar approaches have been commonly applied in the SHE area. For example, root cause analysis (RCA) can be easily applied to accident investigations. One-point lessons can be utilized as a way to communicate key safety lessons and learning. Visual factory and mistake-proofing concepts can be leveraged to create a safer workplace.

1.5 The Value of Safety, Health, and Environment

A key principle of Lean is that organizational resources should be focused on the value chain. All activities must add customer value, and those that do not add value must be considered waste. Devoting organizational resources to SHE adds value from both a humanistic and a business standpoint. Preventing accidents to people and the environment prevents much need-less pain and suffering. The common saying that "Safety is for Life" can be used to highlight the importance of safety. Each letter of the word LIFE can be used as a mental reminder of what really is at stake when safety is on the line.

Life	– The "L" stands for life and reminds me that my life and the ability to enjoy it are at stake when it comes to safety.
I	– The "I" means me and my body parts are in the balance when safety is on the line.
Family	– The "F" stands for family and friends and reminds me that they depend upon me and will suffer if I am injured.
Everything	– The "E" stands for everything, because everything that I enjoy—my ability to engage in exercise, hobbies, and normal everyday activities—is at stake.

In addition to the obvious personal, moral, and humanistic reasons to be concerned about SHE, safety, health, and environmental protection is good business. SHE is about protecting an organization's most important assets: its people, its equipment and facilities, its brands, and ultimately its reputation. In other words, SHE excellence is value enhancing for an orga-nization: protecting people, profit, and public trust. According to the Global Environmental Management Initiative (GEMI) there is "compelling evidence of the link between environmental, health, and safety (EHS) activities and shareholder value…. There is considerable evidence that EHS contributes to shareholder value in a variety of ways—not only through 'tangible' contribu-tions such as risk reduction and profitability improvements, but also through 'intangibles' such as brand equity, human capital and strategic execution."[14]

The belief is that superior management of SHE issues and risks is an indicator of effective overall management, which in turn drives excellent financial performance. Several recent studies and publications have confirmed the belief that good safety, health, and environmental performance provide share value:

- A study by Goldman Sachs showed that among the six industry sectors covered—energy, mining, steel, food, beverages, and media—companies that are considered leaders in implementing environmental, social, and governance (ESG) policies have outperformed the general stock market by 25% since August 2005.[15]
- Analysis of pharmaceutical industry stock performance based upon the EcoValue 21 Rating Index reveals that companies with above average environmental ratings have outperformed companies with below average ratings by approximately 17% (1,700 basis points) since May 2001.[16]
- Another study by Goldman Sachs showed that investors could have had increased returns (25–38%) over the past four years if they had incorporated workplace health and safety measures into their strategy.[17]
- In an extensive literature review by Innovest Strategic Value Advisors, an international investment research firm, found that "good environmental performance can benefit financial performance.... In 85% of the studies assessed, we found a positive correlation between environmental governance and/or events and financial performance."[18]

In recent years public and business attitudes toward safety, health, and environment have undergone a fundamental change. Gone are the days where a "production first, and at any cost" attitude is acceptable corporate behavior. Today it is a strongly held public expectation that businesses will conduct their activities in a safe and environmentally sustainable fashion. Safety, health, and environmental initiatives aren't just good citizenship. They are good business as well. High-performing organizations have a clear understanding of management guru Peter Drucker's contention that "the first duty of business is to survive, and the guiding principle of business economics is not the maximization of profit, it is the avoidance of loss."[19] Consequently, world-class organizations recognize that losses such as accidents and environmental incidents are the key things that a manager can truly manage, and by effectively doing so, can transform top-line dollars (revenue) into bottom-line dollars (profit).

The Global Environmental Management Initiative has developed an excellent model of the SHE value proposition that summarizes the many ways that safety, health, and environmental excellence add value to the organization (see Figure 1.4).[20] According to GEMI there are three pathways in which SHE excellence adds to shareholder value:

1. A direct and tangible pathway
2. A direct and intangible pathway
3. An indirect and intangible pathway

Via pathway 1 the direct tangible benefits of SHE excellence include increased profitability, greater asset utilization, and higher service levels.

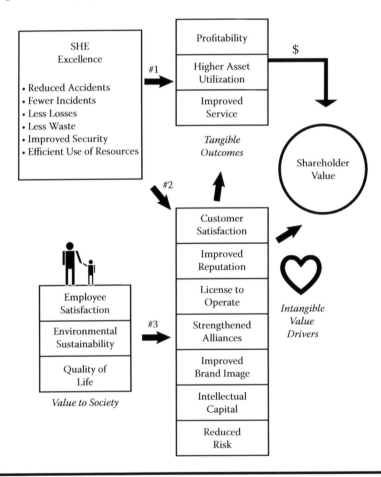

Figure 1.4   The SHE value proposition based upon GEMI figure entitled *Pathways Linking EHS to Shareholder Value*. (From Global Environmental Management Initiative. "Clear Advantage: Building Shareholder Value, Environmental Value to the Investor"; February, 2004, Page 5. www.gemi.org. With permission.)

An effective SHE process results in fewer accidents, environmental incidents, and associated losses, which in turn adds money to the organization's bottom line. In addition, the SHE management system contributes directly to profit by generating revenues from waste utilization, recycling, and greater resource utilization. With fewer accidents and environmental incidents an effective SHE process results in less equipment downtime, resulting in higher asset utilization. As a result of less downtime, the organization achieves the direct benefit of increased service levels.

An effective SHE process also contributes directly to several key intangible shareholder value drivers (pathway 2). SHE excellence results in greater customer satisfaction, improved corporate reputation, positive brand image, beneficial organizational alliances, business continuity, and reduced risk. SHE expertise and specialized knowledge can also benefit the organization's technology, innovation, and business processes. A company can differentiate itself via outstanding SHE performance through enhanced corporate and brand image. Today, there is growing acceptance that these intangible nonfinancial factors have a significant impact on overall share value. In fact, research indicates that more than one-third of a company's market value can be attributed to intangible factors.[21] An effective SHE process positively influences these intangible value drivers.

Finally, as shown in pathway 3, SHE excellence indirectly affects shareholder value by creating value for external stakeholders and society. A leading SHE process contributes to employee satisfaction, environmental sustainability, and overall quality of life. Good SHE performance creates value for stakeholders, thus it favorably affects their opinions and perceptions of the organization, which in turn positively affect intangible shareholder value drivers such as brand image and company reputation.

If one takes a broad and forward-looking view of the safety, health, and environmental arena, it is clear that SHE issues will continue to grow in importance to organizations. Indeed, it can be argued that one SHE issue—climate change and the control of greenhouse gases—is the environmental and business issue of our time and of the future. In fact, in Infor's white paper "Going Green" it is maintained that "the most important driver of business challenges and opportunities in our lifetimes (and our children's) will likely be the Green revolution.... As leading companies embark down the Green Road, they'll transition from a position of being aware of the environmental (health and safety) responsibilities to one that inculcates green as a core value."[22] The editors of the *Harvard Business Review* state that "companies that get their (climate change/greenhouse gas) strategy right will find vast opportunities to both profit and create social good on a global scale."[23] The belief that organizational safety, health, and environmental efforts add value is so widely accepted today, it can confidently be said that SHE is a societal value. Smart organizations today understand that it is possible to do well by doing good.

1.6 Zero Accident and Incident Vision

The vision of Lean/TPM is to achieve zero losses, and an integral part of this vision is zero accidents and zero environmental incidents. Lean and the SHE pillar are about creating a world-class supply chain and organization, and sharing a common vision of excellence—a vision characterized by zero breakdowns, zero defects, zero accidents, zero incidents, and ultimately zero losses. Both Lean and SHE aim to eliminate losses through employee involvement and the application of systematic continual improvement techniques. By utilizing continuous improvement tools and techniques the SHE pillar aims to establish a safer man–machine–material (3 M) interface that ultimately achieves zero injuries, zero illnesses, and zero environmental insults. The key approach that the SHE pillar employs to realize this vision of SHE excellence is the systematic, proactive removal of risk. The SHE pillar focuses on developing a management system that creates safe equipment and safety-conscious people.[24]

Upon first glance, achieving zero accidents and environmental incidents may appear patently absurd and virtually impossible for many organizations. It must be understood, however, that the journey toward zero is taken one step at a time. Achieving "zero" simply requires each person to work safely one task, one day at a time. In fact, most people have the ability to work safely without harming themselves or the environment most of the time.

Even in poor safety performing organizations, with an annual OSHA recordable rate of 5, approximately 95% of employees work accident-free during a work year. The challenge of the SHE pillar is to extend the length of time each employee can work accident- and incident-free by giving them the knowledge, skills, ability, and motivation to systematically remove risk from their workplace.

Zero is the right safety, health, and environmental vision. Accepting accidents and environmental incidents as the normal course of doing business is both a financially and morally questionable position. Accidents and their associated losses can be costly to an organization's bottom line. According to the National Safety Council the total cost of on-the-job injuries in the United States was a staggering $156.2 billion in 2003, with the average cost of a disabling injury being $38,000.[25] Because many companies are self-insured, these accident costs go directly to the organization's bottom line. In fact, if a company has a profit margin of 10%, $380,000 in sales are required to offset the costs of an average disabling injury. In light of the high cost of accidents, it is not surprising that, according to a survey conducted by Liberty Mutual Insurance, 95% of business executives report that workplace safety has a positive impact on their company's financial performance, and 61% of the executives believe that their companies receive a return on investment (ROI) of more than $3 for every $1 that they invest in safety.[26] Zero accidents and zero incidents is the right goal, for SHE excellence makes good business sense.

It becomes clear that aiming for a 100% safe work environment is the appropriate goal from a humanistic standpoint, when one understands what accepting 99% safe really means:

- A 99% safe standard for airport takeoffs and landings equates to approximately 27 aircraft accidents per day at Atlanta's Hartsfield Airport.[27]
- A 99% safe standard for births in United States means over 42,000 accidents during delivery each year.[28]

These examples clearly demonstrate that a 99% safe standard for airports and hospitals is not acceptable. We demand a higher level of safety performance from these industries, feel that this lofty goal is reasonable, and expect them to meet it. Just as we routinely demand high levels of safety performance from others, we should also demand it of ourselves and our own organizations. From a human perspective, zero accidents and incidents is the right goal and it can be achieved.

Realizing the vision of zero accidents and environmental incidents is not easy, but it is not impossible either. In the 1960s many people said that President John F. Kennedy's vision of sending a man safely to the moon and back within a decade was impossible. Critics said that it was a foolish and unsafe goal. The naysayers argued that we didn't have rockets powerful enough to get a man there and back, we didn't have computers capable of calculating a flight path, and government scientists weren't smart enough to plan the mission. Despite this, NASA rallied behind President Kennedy's vision and through the efforts of many people, NASA was able to make the dream of putting a man on the moon and returning him safely to Earth a reality. Likewise in the area of safety, great things can be accomplished through the combined efforts of many individuals. Achieving zero accidents and zero environmental incidents is possible if everyone in an organization works together to remove and control risk.

Safety is an essential part of doing any task or job well. If one performs a difficult task and injures or kills someone in the process, the task is rightly deemed a failure. For example, if the United States had successfully landed men on the moon but did not return them safely to Earth, the entire mission would have been deemed a failure. In most people's minds even the impressive technological feat of a lunar mission is greatly diminished if it is not done safely. Likewise, notable production and industrial achievements are largely devalued by safety, health, and environmental incidents and losses. It can be argued that the success of the U.S. space program is due, in large part, to its emphasis on safety. The attention to detail and operational discipline required to achieve a goal of zero accidents had a positive effect on the overall management of the space program. In fact, reversals and setbacks in the space program have generally occurred when NASA's focus on SHE excellence has been diminished.

1.7 Strategy for Eliminating Accidents and Environmental Incidents

Unfortunately, there is no single magic bullet that can eliminate all accidents and incidents. If there were one ingredient that could guarantee SHE success, many organizations would already be implementing it and zero accidents would be commonplace. Realizing the "zero vision" requires all parts of an organization to do many things well. SHE excellence requires a company to focus on the three key areas of organizational design and functioning, known as the 3 Ms: man, machines, and management systems. Addressing only one of the Ms, such as machines, will only deliver limited SHE success.

From a simplistic perspective, as indicated in Equations (1.1) and (1.2), accidents are a function of hazards and exposure, and risk is a function of the severity of the hazard and the amount of exposure.

Equation (1.1) The Accident Equation

$$\text{Accidents} = f\,(\text{Hazard}) + (\text{Exposure})$$

Equation (1.2) The Risk Equation

$$Risk = f \text{ (Severity of Hazard + Degree of Exposure)}$$

For an accident to occur two things need to happen:

1. A hazard needs to exist.
2. A person needs to be exposed to the hazard.

These safety equations tell us that a accident cannot occur in the absence of hazards. An accident is impossible when one is in a 100% safe environment, exposed to no hazards.

   Although a hazard-free workplace is a utopian situation that is rarely encountered, the hazard component of safety equation (1.3) provides a clue to one of the key approaches to achieving safety excellence: hazard control.

Equation (1.3) The Safety Equation

$$Safety = f \text{ (Hazard Control) + (Exposure Control)}$$

Hazard control is a common tactic used by many organizations to reduce accidents and improve safety. Hazard control can be considered an "engineering model for improved safety" because the focus is primarily on removing hazards from the workplace through improved equipment and process design, maintenance, and repair. Risk is reduced by eliminating the hazard or reducing its severity. Traditional safety programs have a focus on "things" rather than people, thus they commonly devote significant resources to hazard control in the workplace.

   The safety equation tells us that another way to ensure safety is via exposure control: the elimination of all potential exposure to hazards. A hazard that no one is exposed to will not cause an injury or illness. For example, if a hazardous material, such as chlorine, is handled according to procedures in a completely enclosed, fail-safe system that eliminates all potential for exposure, accidents can be avoided. Without exposure, a person cannot become a target of a hazard, and there can be no accident. Although the hazard is still present, an accident does not occur because no one is exposed to the hazard. Exposure control can be considered the "medical model for improved safety" inasmuch as it relies on preventing people from becoming a target of things that are bad for them, or can harm them. Just as a doctor provides advice to a patient on how to reduce exposure to things that are unhealthy, this approach focuses on preventing worker exposure to things that are unsafe. Risk is reduced by reducing the degree of exposure. Safety programs often devote some resources to exposure control by means of engineering controls, personal protective equipment, and employee training. Many safety programs, however, are less effective at providing workers with the skills and ability to prevent themselves from becoming the target of hazards via their own risky work practices and actions. People-centered safety and behavior-based safety are processes that provide workers with the ability to work safely and avoid exposure to hazards. Workers can work safely in the presence of a hazard if they have the knowledge and skills to effectively avoid the hazard.

Organizations can reduce accidents by controlling hazards and reducing exposure to them. As indicated in Figure 1.5, organizations typically focus first on systematically removing risk from their machines and equipment, their so-called "SHE hardware." By improving the design, reliability, and inherent safety of plant equipment many accidents and environmental incidents can be avoided. Typically, SHE activities implemented by organizations concerned about machine-related risks include equipment design reviews, 5S audits, and physical condition inspections. Focusing solely on equipment hazards, however, will not achieve SHE excellence. Addressing only machine-related safety issues achieves some success, but over time the organization's safety performance will tend to plateau. To break through this

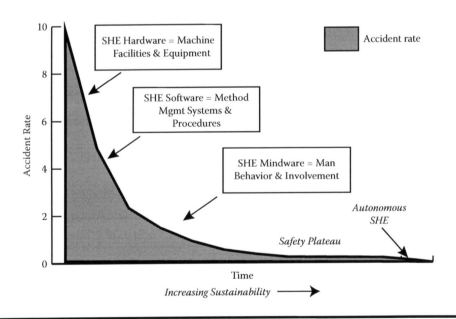

Figure 1.5   The components of safety, health, and environmental excellence.

performance plateau an organization's SHE process also needs to consider and address people and management systems.

Implementation of management systems, the "SHE software" that defines an organization's ways of working, is often the second area that an organization tends to address in the journey toward SHE excellence. Commonly, as a SHE process matures it becomes clear that establishment of a comprehensive SHE management system is required for the organization to systematically and efficiently remove SHE risks from the workplace. Without a structured SHE management system, variation in the way SHE risks are addressed is likely to occur, resulting in accidents and incidents. A well-executed SHE management system enables an organization to effectively identify and control its significant risks by defining the procedures, methods, and operational controls that will be employed to address these risks. In other words, SHE management systems can be considered as enabling systems that make an organization's hazard and exposure control more effective. Many leading organizations implement comprehensive SHE management systems based upon ISO 14001 and OHSAS 18001 standards.

After addressing "SHE hardware" and "SHE software," the journey toward SHE excellence continues with the control of risky work practices or behavior, known as "safety mindware." Ultimately all successful SHE processes have an emphasis on people rather than things. Safety mindware initiatives frequently involve the implementation of processes that provide employees with the knowledge, skills, and motivation to work safely, such as hands-on training activities and formal SHE observation and feedback systems. SHE observation and feedback processes prevent accidents by (1) creating awareness that at-risk behaviors affect personal safety, (2) increasing knowledge of safe work practices, (3) establishing an atmosphere that encourages all personnel to work safely, and (4) removing barriers to safe work. A formal SHE observation and feedback process works via the following mechanisms:

1. *Training:* Training provides information and instruction that increases awareness of at-risk behaviors, and techniques for increasing personal safety.
2. *Accident Analysis:* Behavioral analysis of accidents increases understanding of what behaviors are key to safety at the site.
3. *Observation and Feedback:* Observation and feedback provide opportunities for individual learning and positive reinforcement for safe actions, and negative or corrective reinforcement for risky behaviors committed by individual coworkers.

4. *Charting:* Charting provides visual information that serves to educate site personnel on the findings of the observation process. This information can be used by personnel to take actions to increase personal safety.

5. *Group Meetings:* These serve as another mechanism for communicating learning derived from the observation process and to encourage safe work.

6. *Implementing Corrective Action:* Formal problem-solving based upon the observation data enables targeted solutions to be implemented for specific safety issues that have been identified.

It is only by addressing SHE hardware, SHE software, and SHE mindware that sustained safety excellence can be achieved. Organizations whose SHE process addresses only one or two of these areas are likely to experience limited success and ultimately a leveling (plateau) in their SHE performance. A well-designed SHE pillar considers these three components of SHE excellence.

Endnotes

1. Ventana Research (2007). Strategies to run a Lean supply chain. How principles of Lean manufacturing transfer benefits to operations. Ventana Research White Paper. San Mateo, CA.

2. Womack, James and Daniel Jones (1996). *Lean Thinking: Banish Waste and Create Wealth in Your Corporation. Part 1: Lean Principles.* Simon & Schuster, New York.

3. Ford, Henry (1922). *My Life and Work.* Nevins and Hill. Reprinted by Greenbook, LaVergne, TN, (2010), p. 49.

4. Womack and Jones, op. cit., pp. 22–23.

5. Ford, Henry, op. cit., pp. 69–76.

6. Liker, Jeffrey (2004). *The Toyota Way. 14 Management Principles from the World's Greatest Manufacturer.* McGraw-Hill, New York, p. 10.

7. Liker, op. cit., pp. 1–34.

8. Ohno, Taiichi in "High quality with safety: Kanban and just-in-time," unpublished internal Toyota document referenced in Liker, Jeffrey (2004). *The Toyota Way. 14 Management Principles from the World's Greatest Manufacturer.* McGraw-Hill, New York, p. 34.

9. Womack, James et al. (1990). *The Machine That Changed the World: The Story of Lean Production.* Rawson Associates, Division of Macmillan, New York, p. 13.

10. Womack et al., op. cit., pp. 12–13.

11. Shirosi, Kunio (1996). *Total Productive Maintenance: New Implementation Program in Fabrication and Assembly Industries*. Japan Institute of Plant Maintenance (JIPM), pp. 2–21.
12. Maslow, Abraham H. (1943). A theory of human motivation. *Psychological Review*, 50: 370–396.
13. Shirosi, op. cit., pp. 500–501.
14. Thomas, Jim and John Harris (February 2004). "Clear advantage: Building shareholder value. Environment value to the investor." GEMI (Global Environmental Management Initiative). p. 2. Retrieved from: http://www.gemi.org/resources/GEMI%20Clear%20Advantage.pdf
15. Alderton, Margo (2007). "Green is gold, according to Goldman Sachs study." *CRO Corporate Responsibility Officer*. June 2007. Retrieved from: http://www.thecro.com/node/490. Original paper: Anthony Ling et al. Goldman Sachs Introducing GS Sustain, June 22, 2007 at: http://www.unglobalcompact.org/docs/summit2007/gs_esg_embargoed_until030707pdf
16. Baue, William (2002). "Eco-efficient pharmaceutical companies have higher share value." *Social Funds*. July 03, 2002. Retrieved from: http://www.social-funds.com/news/article.cgi/873.html. Original paper: Innovest ecovalue 21 study: The global pharmaceutical industry uncovering hidden value potential for strategic investors.
17. Goldman Sachs JBWere (October, 2007). "Media Release. Goldman Sachs finds valuation links in workplace health and safety data." Retrieved from: http://www.gsjbw.com/documents/About/MediaRoom/GSJBW-WHS-Report-Media-Release.pdf
18. White, Ander and Matthew Kiernan (September 2004). Innovest Strategic Value Advisors. Corporate environmental governance: A study into the influence of environmental governance and financial performance. Environmental Agency, Bristol, UK.
19. Drucker, Peter (1972). *Technology, Management, and Society*. Harper and Row, New York.
20. GEMI—Global Environmental Management Initiative. (February 2004). "Clear advantage: Building shareholder value, environment: Value to the investor." p. 5. Retrieved from: http://www.gemi.org/resources/GEMI%20Clear%20Advantage.pdf
21. Low, Johnathan and Pamala Cohen Kalafut (2002). *Invisible Advantage: How Intangibles Are Driving Business Performance*. Persius Press, Basic, Cambridge, MA.
22. Infor (2007). "Going green: Practice for a profitable future." Alpharetta, GA. Retrieved from http://www.infor.com/goinggreen/
23. *Harvard Business Review* (October, 2007). Climate business, business climate. Forethought Special Report.
24. Suzuki, Tokutaro (1994). *TPM in Process Industries*. Productivity Press, New York, p. 327.

25. National Safety Council (2005). *Injury Facts 2004 Edition.* NSC Press, Chicago.
26. Liberty Mutual (2001). "A majority of U.S. businesses report workplace safety delivers a return on investment." Liberty Mutual News Release. Boston. August 28. Retrieved from: www.libertymutual.com
27. City of Atlanta (2007). Hartsfield Airport Fact Sheet. Atlanta. Feb. 2007. Retrieved from: http://www.atlanta-airport.com/Passenger/pdf/Fact_Sheet_2010.pdf
28. National Center for Health Statistics (2006). "Fast Stats A to Z." CDC. Atlanta. Dec. 2006. Retrieved from: http://www.cdc.gov/nchs/fastats/births.htm

Chapter 2

Autonomous SHE

The safety of the people shall be the highest law.

Marcus Tullius Cicero
Ancient Roman statesman, 106–43 BC

2.1 The Path toward Autonomous Safety/Autonomous SHE

A safety culture is not a program that can be acquired and simply bolted onto an existing organization. It is not a simple process input that can be supplied whenever needed. Instead, safety culture is the complex outcome of the values, beliefs, norms, and numerous actions and interactions of organizational members over time. A safety culture lies at the core of an organization, and is an enduring fundamental force that influences how organizational members think, behave, and perform their work. Analysis of several well-publicized catastrophic accidents has indicated that a weak safety culture, or lack of a safety culture, is a key underlying cause of safety failures. For example, the study of NASA's *Challenger* explosion revealed that it was not simply the result of a technical O-ring failure, but rather the outcome of a dysfunctional organizational safety culture that resulted in faulty decision making and the downplaying of risk.[1] The 1988 explosion and resulting fire on Occidental Petroleum's North Sea Piper Alpha offshore oil rig, which tragically killed 167 people, has also been attributed to breakdowns in organizational safety culture. Pate-Cornell's accident analysis of this disaster reveals that it was not a random act of God, but instead a "self-inflicted" tragedy resulting from a series of management errors that were fostered by a dysfunctional safety culture.[2] More recently it has been argued that British Petroleum's recurring history of catastrophic safety failures including refinery explosions, Alaskan oil pipeline spills, and the explosion and massive oil spill of the Deepwater Horizon oil rig in the Gulf of Mexico are symptomatic of a woefully deficient safety culture. The U.S. Chemical Safety Board and the Baker Safety Review Panel indicated that BP had "a corporate safety culture that may have tolerated serious and longstanding deviations from good safety practice."[3]

Building a strong safety culture takes consistent and sustained organizational effort, but once established it is a key determinant in the organization's safety success and overall performance. In other words, the journey toward SHE excellence involves a dramatic transformation of an organization's culture that not only improves safety, health, and environmental performance, but also the overall organizational functioning. A well-developed safety culture is a determining factor in the safety performance of a wide variety of organizations from the petrochemical industry to healthcare.[4]

As an organizational safety, health, and environmental culture evolves and matures it tends to go through several distinct phases that improve the functioning of the enterprise's SHE systems and result in progressively lower accident and incident rates. DuPont[5] and Hudson[6] have developed safety culture continua or maturity models that describe some of the key steps in the evolution of an organizational safety culture. Figure 2.1 provides another perspective on the evolution of an organizational safety, health, and environmental culture that proposes that higher levels of SHE culture maturity and sustained levels of outstanding SHE performance can be realized by incorporating and leveraging Lean principles. The initial stage in this SHE culture continuum is termed the "uncaring culture." In this pathological and reactive stage, safety, health, and environmental management is neither an organizational value nor a priority. Organizations at this stage have little interest in workplace safety, health, and environmental issues, seeing protection of employees and the environment as a business burden and unnecessary cost. Typically the business places a lower value on people and views labor as an expendable resource that can simply be replaced when accidents happen. Ironically, as the organization worries about the cost of SHE programs,

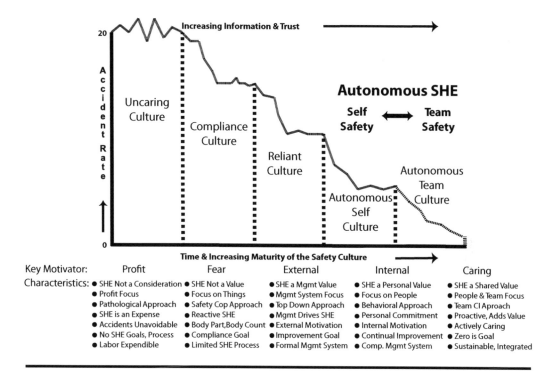

Figure 2.1   Stages in the evolution of a safety culture: the path toward autonomous SHE (expansion of DuPont's[5] safety culture model, reprinted with permission).

significant costs are incurred from repeated safety, health, and environmental incidents and losses. At this primitive stage, SHE matters are not on the management agenda and virtually no organizational resources are devoted to addressing workplace risk. It is not surprising that during this early stage in cultural development no lasting improvement in SHE performance is realized and accidents rates tend to exhibit normal variation at very high levels. Fortunately, the SHE cultures in most modern organizations have evolved beyond this primitive and callous stage.

The second phase in SHE culture maturity is the "compliance culture." At this early stage SHE is not a shared value, and the primary reason the organization devotes any resources to SHE is to avoid regulatory citations and penalties. As a result, the organization's limited SHE process tends to be reactive in nature focusing at the top of the accident pyramid. The "body count, body part" method of SHE management is commonly employed in which lagging indicators, such as the number of fatalities and the number of injuries, are used to determine future actions. The primary focus of the safety process is on things or physical conditions. After an accident occurs, investigations are launched, and reactive actions are taken primarily to address the condition-based causes of the accident. There is limited involvement of management and employees in the SHE process, and the organizational culture is characterized by limited sharing of safety related information and a lack of trust. Typically, the site SHE professional, if there is one, is the only person in the enterprise concerned about SHE and devoting any effort to this area. The organization's rudimentary compliance-based SHE process, focused primary on safety conditions or hardware, has some initial success in reducing workplace accidents and incidents, however, accident rates soon plateau at a relative high level.

The third stage in the evolution of a SHE process is known as the "Reliant Culture," since the organization relies primarily on management's efforts and the external motivation of employees to achieve workplace safety. In this phase, SHE is a common management value and leadership uses a top down approach to implementing safety, health, and environmental requirements and programs. Modest SHE success is achieved via supervisory and management driven activities. Since employees are often considered the source of safety problems; improvement is achieved via the establishment of rules, their enforcement, and discipline of violators. At this stage there is increased focus on building a formal safety, health, and environmental management system. However, since there is limited employee involvement in the SHE process, limited SHE success is achieved.

The next phase in SHE culture maturity is termed the "autonomous SHE self culture" since employees have accepted personal responsibility for SHE; and take independent, proactive action to ensure their safety and protection of the environment. As the SHE culture matures there is increasing employee involvement in the SHE process, and growing commitment to SHE excellence. Employees make safety a personal value, and are internally motivated to work safely because they understand that they are the primary beneficiaries of a safe workplace. Coworkers accept that their work practices and actions influence their own safety, and work to manage their own risky behavior. During this stage there is an increasing focus on the behavioral causes of accidents and incidents, the so called "safety mind-ware." Lean techniques and TPM methodologies are utilized to enable employees to develop the knowledge and skills to work safely on their own. In an "autonomous maintenance culture" employees take ownership of their equipment and learn how to operate it efficiently, while in an "autonomous safety culture" employees take ownership of their safety and learn how to work without injury or incident. As employees take ownership for safety, health, and environment, and become actively involved in the organization's SHE process; there is a corresponding reduction in the organization's accident and incident rates. Even in this advanced stage of safety culture, SHE performance tends to plateau and the vision of zero accidents, zero incidents, and zero losses is not realized.

During the final phase in the evolution of a SHE culture, known as the "autonomous team culture," all elements are present to make the vision of triple zero (zero accidents, zero incidents, and zero losses) a reality. In this mature cultural stage, the organization's safety, health, and environmental processes are fully integrated into its decision making, management systems, and ways of working. The enterprise effectively manages the hardware (physical conditions), software (management systems), and mind-ware (behavioral aspects) of safety. At this stage there is a perfect integration of individual independent safety action, and interdependent team safety action. Coworkers are able to work safely on their own, but also use teamwork to achieve higher levels of safety, health, and environmental performance. SHE is seen as a team game where everyone plays an important role, and everyone wins. Employees routinely and freely provide safety support, feedback, and information to each other in order to prevent injury or environmental harm. Coworkers have developed the ability to work safely in teams, and to work together to implement safety *kaizens* or improvements. The safety culture is characterized by high levels of trust, actively caring for others, and cooperation. SHE excellence is seen as adding value to the business, and Lean methodologies are leveraged to achieve sustained levels of outstanding SHE performance.

Under a well-established and successful SHE pillar, the entire organization strives to achieve the state of "autonomous safety," where there is an optimal blending of:

1. Proactive personal and group SHE action
2. Individual and team-based continual improvement in safety, health, and environment
3. SHE support provided by individuals and teams
4. SHE activities designed to address safety hardware, software, and mind-ware.

At this advanced stage, employees are able to work safely independently and in collaboration. This journey toward "autonomous safety" moves the organization toward sustained world class SHE performance in which zero accidents, zero environmental incidents, and zero losses become a reality. The organization transforms from a reactive or dependent culture to a high performing safety culture where all individuals make "safe behavior" a critical part of their way of thinking and way of working. Individuals have the knowledge, skills, and motivation to work safely on their own, and in teams.

Coworkers routinely provide safety feedback and support, and take proactive action to improve the SHE management system, and overall SHE performance. In an "autonomous safety culture" it is common for coworkers to remind each other about the group's agreed and required safety norms and safety behaviors. Because safety is seen as a critical part of good job performance it is a normal part of the workplace conversation. The entire organizational culture and work environment encourages, fosters, and values safe work practices. The SHE process is owned by all employees, and everyone from top management to line workers are actively involved in meaningful SHE work. Safety, health, and environmental excellence is seen as bringing value to the organization by not only reducing losses, but by improving overall operational efficiency and effectiveness. From the boardroom to the shop floor SHE is an unalterable value and everyone truly believes that "safety pays." The entire organization works together to achieve sustained, continual improvement in SHE performance and to realize the goal of triple zero.

As depicted in Figure 2.2, the journey toward an autonomous safety culture involves several key transformational steps for an organization, including the following:

1. Making SHE an organizational value
2. Seamlessly integrating SHE into the organization's way of working
3. Establishing a robust SHE management system
4. Creating a sustained SHE culture

The first step toward autonomous safety focuses on making SHE an organizational value, and therefore involves several initiatives designed to change the mindset of leadership and employees. Workshops on SHE leadership are conducted in order to transform managers into committed and involved SHE leaders. Senior managers receive safety coaching and are expected to develop personal safety action plans and to demonstrate visible safety leadership. DuPont maintains that management's heartfelt display of safety commitment and engagement, which they call "Felt Leadership,"[7] builds trust and credibility and is the foundation of safety excellence. Values and envisioning exercises are conducted in order to clearly define the organization's vision of SHE excellence and the values, behaviors, and attributes needed to

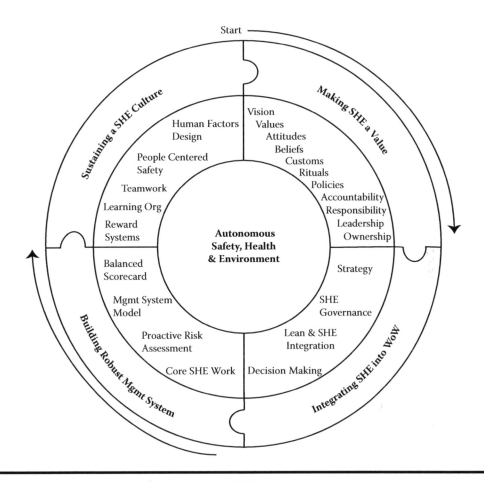

Figure 2.2 Journey toward autonomous SHE.

achieve it. Organizational vision, mission, and policies are updated to reflect the organization's commitment to sustained safety, health, and environmental excellence. In addition, the business rethinks its customs and rituals to ensure that they are consistent with and supportive of its SHE values. Upon completion of this initial phase employees take personal responsibility, accountability, and ownership for workplace SHE and begin to assume leadership for safety in their work areas.

The second step in the quest to establish an autonomous safety culture involves formally integrating safety, health, and environmental issues into the business's way of work or WoW. Transformation of the organization's WoW entails aligning SHE goals and corporate strategy, establishing a strong SHE governance model, and incorporating SHE into all organizational decision-making processes. Systems are established to ensure that SHE is not managed separately from the organizational strategy and decision making, but rather is considered a key factor and determinant in business decisions and is leveraged for competitive advantage. In Lean organizations SHE is seamlessly assimilated into each employee's work practices by means of the autonomous maintenance pillar. As workers learn about and take ownership of their equipment, they simultaneously learn about equipment and task hazards and take ownership of workplace safety. As the Lean process increases employee motivation, job knowledge, and task competency it also increases safety motivation, safety expertise, and safety skills.

The third step requires the implementation of a comprehensive SHE management system that proactively focuses resources on significant SHE risks and potential losses. This phase recognizes that because many accidents and incidents ultimately result from a management system breakdown, it is essential to design a robust management system that has adequate checks and balances that enable the quick and reliable identification and correction of management failings. The management system aligns SHE targets with the organization's key safety, health, and environmental risks and requirements, and ensures that adequate resources are made available to achieve the SHE objectives. Tools for building a leading SHE management system include the following:

1. A SHE balanced scorecard that clearly defines safety, health, and environmental targets and objectives
2. A management system model such as ISO 14001 that gives structure to the SHE process
3. A formal core SHE process that puts the management system into action by defining the routine SHE work that is required to achieve SHE success

In the final phase, a sustainable safety culture is established where SHE is a shared value that permeates all decisions and activities. There is true integration of safety, health, and environment into the overall management system and strategic decision making. The reward system reinforces the organization's safety values and desired behaviors. People-centered safety and behavior-based safety are leveraged to promote safe work practices, and human factors design is utilized to compensate for predictable human weaknesses and errors. The safety system is designed to foster safe work; however, if mistakes are made, the system mitigates the adverse effects of errors. Consistent with James Reason's Swiss cheese model of accident causation,[8] the organization's multiple slices of cheese or layers of safety defenses, barriers, and safeguards are so complex and complete that weaknesses or holes in the system are limited; therefore, when errors do occur they are controlled and the losses are prevented. Finally, the mature and advanced safety culture is characterized by trust, teamwork, and sharing of information. As a learning and adaptive organization it is "skilled at acquiring and transferring knowledge and at modifying its behavior to reflect new knowledge and insights."[9]

2.2 Proactively Removing SHE Risk

Contrary to popular opinion, safety is not common sense. Rather, safety, health, and environmental protection are the result of the proactive, careful, and systematic assessment and management of risk. SHE risk assessment requires data and specialized skills, not common sense. Indeed, the use of so-called common sense, without adequate risk assessment, results in numerous injuries and illnesses every year in common everyday activities such as operating motor vehicles and power tools. Numerous tragic automobile accidents occur because people use so-called "common sense" while performing the highly hazardous task of operating a vehicle traveling 60 miles an hour on a crowded highway.

SHE success is achieved the old-fashioned way, by doing work! Just any SHE work, however, will not suffice; the work must be the right work: activities that are directed toward identifying and removing risk from the work environment. Although most organizations conduct SHE activities or work, they are often "meaningless rituals" performed without data and without a focus on the removal of risk and the elimination of loss. As shown in Figure 2.3, avoiding accidents and incidents requires an organization to proactively remove risk by systematically analyzing and acting upon data at the bottom of the accident or safety pyramid—data related to unsafe or substandard conditions and at-risk behavior. An effective SHE management system has processes in place to systematically identify and remove both at-risk conditions and at-risk behavior.

Another useful way to analyze an organization's SHE risks is to develop a SHE loss tree, shown in Figure 2.4, that details the different types of safety, health, and environmental losses suffered by an organization. In light of the management axiom, "What gets measured, gets done; and what gets measured and rewarded, gets done well," a well-developed loss tree tracks both the number and cost of each type of loss. In this way, an organization can manage both the frequency and severity of loss events.

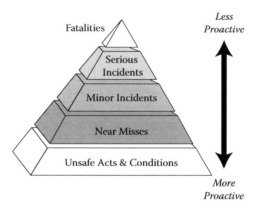

Figure 2.3   The safety pyramid.

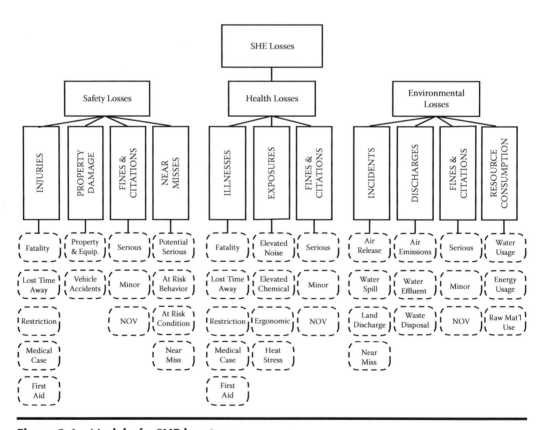

Figure 2.4   Model of a SHE loss tree.

2.3 Integrating the SHE Pillar with the Other Pillars

Safety, health, and environment should not be managed differently from other business issues, or separately, outside the organization's management system. In order to harness the full power of the Lean/TPM continuous improvement methodology and to integrate safety, health, and environmental issues more effectively into an organization's culture and management system, efforts must be made to incorporate SHE into all of the continuous improvement pillars. The SHE pillar team should promote the implementation of specific SHE improvement initiatives for each Lean/TPM pillar. These improvement initiatives should be designed to fully leverage the continuous improvement tools and techniques of each pillar to improve SHE performance.

Examples of some specific SHE initiatives that can be implemented for the various pillars include the following:

Autonomous Maintenance Pillar
1. SHE tagging
2. SHE visual controls
3. Operator SHE checklists
4. SHE standards
5. SHE inspection points

Preventative Maintenance Pillar
1. Lockout/tagout, zero energy state (LOTO/ZES)
2. SHE permit systems

Focused Improvement Pillar
1. SHE *kaizens*
2. SHE root cause analysis (RCA)
3. 6S for SHE (physical condition *kaizens*)
4. People-centered safety (at-risk behavior *kaizens*)

Training and Education Pillar
1. SHE training plan
2. SHE one-point lessons (OPLs)
3. Application of *shu ha ri*
4. *Kaizen* gym

Early Management Pillar
1. Hazards and operability reviews
2. Fault tree analysis (FTA), failure mode effects analysis (FMEA)
3. Project SHE checklists
4. 30–60–90 day reviews

Endnotes

1. Vaughan, Diane (1996). *The Challenger Launch Decision: Risky Technology, Culture, and Deviance at NASA*. The University of Chicago Press, IL.
2. Pate-Cornell, M. Elisabeth (1993). Learning from the Piper Alpha accident: A postmortem analysis of technical and organizational factors. *Risk Analysis*, 13: 2.
3. Baker, James et al. (2007). "The report of the BP U.S. refiners independent safety review panel. P 41." Retrieved on Sept. 1, 2010 from: http://www. bp.com/liveassets/bp_internet/globalbp/globalbp_uk_english/SP/STAGING/ local_assets/assets/pdfs/Baker_panel_report.pdf
4. University of Michigan (2002). "Patient safety toolkit. University of Michigan Health System." Retrieved on Sept. 1, 2010 from: http://www.med.umich.edu/ patientsafetytoolkit/culture/chapter.pdfhttp://www.aiche.org/uploadedFiles/ CCPS/Resources/KnowledgeBase/Whats_at_stake_Rev1.pdf
5. Whitman, Mark (Nov. 2005). "The culture of safety: No one gets hurt today." *The Police Chief*. 72: 11, Retrieved on Sept. 15, 2010 from: http://www.policechiefmagazine.org/magazine/index. cfm?fuseaction=display_arch&article_id=737&issue_id=112005
6. Hudson, P. (2001). Safety management and safety culture: The long, hard and winding road. Occupational Health and Safety Management Systems. Proceedings of the First National Conference. Crown Content. Melbourne, Australia.
7. Schweitzer, Melodie A. (Nov. 2007). Creating a safety culture through felt leadership. *Industrial Hygiene News*. 13. Retrieved on Sept. 15, 2010 from: http:// www.rimbach.com/scripts/article/IHN/Number.idc?Number=113
8. Reason, James (1988). Achieving a safety culture: Theory and practice. *Work & Stress*. 12, 3: 293–306.
9. Garvin D. (1993). Building a learning organization. *Harvard Business Review*, (July/August): 78–91.

Chapter 3

Preparation Phase: Laying the Foundation for SHE Excellence

The general who wins the battle makes many calculations in his temple before the battle is fought. The general who loses makes but few calculations beforehand.

Sun Tzu[1]

3.1 Management Commitment and Living Leadership

The noted Greek philosopher Plato observed that "the beginning is the most important part of the work."[2] This maxim is true when implementing the Lean SHE pillar, for the initial planning or preparation stage is crucial to the ultimate success of the process. Although the initial inclination is often to rush into the implementation phase of Lean, it is vital to build a strong foundation for the process first. A good starting point for a Lean process and the SHE pillar is obtaining full management commitment to the initiative. Management leadership, and commitment to and visible support for SHE are essential because management sets the vision, values, and priorities for the organization. DuPont argues that senior leadership must demonstrate "felt" safety leadership, meaning they make safety a personal value and act upon that value. According to DuPont, "Felt leadership is a public proclamation of an organization's commitment to caring about people. It is a building block in constructing trust and real-world relationships among employees, customers, shareholders and communities."[3] DuPont's 10 principles of safety leadership should guide the behavior of senior management as an organization implements a Lean SHE pillar process. Leaders must

- Be visible to the organization
- Be relentless about time with people
- Recognize their role as teacher/trainer
- Develop their own safety skills and pass them along to the organization
- Behave and lead as they desire others to do
- Maintain a self-safety focus
- Confirm and reconfirm safety as the No. 1 value
- Place continuous emphasis on and clarity around safety expectations
- Show a passion for ZERO injuries, illnesses, and incidents
- Celebrate and recognize ZERO successes

Another way to look at the SHE leadership that must be demonstrated by senior management is via the "3 H heart, head, and hands model" shown in Figure 3.1. To be truly effective, leaders must feel commitment to safety, health, and environmental excellence in their hearts, believe it in their heads, and act upon it with their hands. In other words, organizational leaders must hold safety, health, and environmental excellence as a personal value, must have an intellectual and emotional commitment to ensuring a safe workplace, and must take personal action to achieve outstanding SHE performance. A leader's commitment to SHE excellence must be unwavering and uncompromising, in good times and in bad. It is said that "the only safe ship in a storm is leadership,"[4] and this must be true for organizational SHE leadership. Senior management should provide a model for the organization

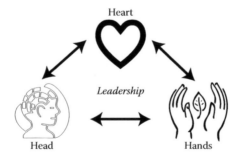

Figure 3.1   The 3 H leadership model.

of living leadership in the SHE area. SHE leaders must practice the seven "Cs" of safety:

- *Care:* Leaders care about people and make SHE a personal and organizational value.
- *Codify:* Leaders make SHE an official organizational value via written policies.
- *Communicate:* Leaders communicate the SHE vision and expectations.
- *Confirm:* Leaders confirm the understanding of all to the SHE vision and values.
- *Commit:* Leaders make a personal and group vow to SHE excellence.
- *Concrete Action:* Leaders live the SHE vision and values via personal example and actions.
- *Cultivate:* Leaders create and cultivate a SHE culture that values and rewards excellence.

Practical ways for senior executives to demonstrate living SHE leadership during the initial preparation phase include

- Issuing a management communication demonstrating personal support for the Lean SHE pillar initiative
- Ensuring that the SHE pillar process is adequately resourced
- Personally attending initial Lean, SHE leadership, and SHE pillar training
- Hosting a Lean and SHE pillar kick-off session
- Endorsing a vision and values statement that links SHE excellence with organizational values

Although the time required to complete the preparation phase of a Lean SHE process will vary depending upon the size and complexity of the organization, three to six months is typically needed to complete all the activities required to lay a strong foundation. Organizations should devote sufficient resources to complete these activities within six months. Taking a longer time to complete the preparation phase of Lean is discouraged because delays tend to result in deterioration of management commitment, decreased employee enthusiasm, and a loss of process momentum. Figure 3.2 lists the recommended SHE activities that organizations should consider implementing during the preparation phase of Lean:

- SHE climate analysis
- Leadership SHE workshop

- Seven Cs of SHE leadership
- SHE vision and values statement
- Zero accident education
- SHE loss tree
- Establishment of a SHE governance model and SHE master plan
- SHE targets and SHE balanced scorecard
- Six S or 5S for SHE
- Lockout–tagout/zero energy state (LOTO/ZES)

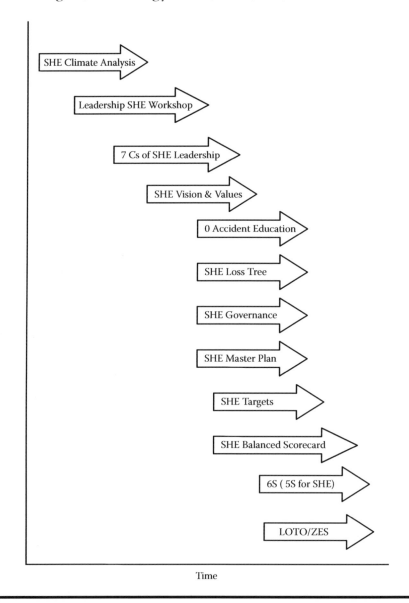

Figure 3.2 Key SHE activities undertaken during the Lean preparation phase.

3.2 Assessing Organizational Readiness

Prior to launching a full-scale Lean SHE pillar process it is important to gauge the organization's readiness for the undertaking, and to determine whether the existing organizational values, climate, and culture are supportive of SHE excellence. Unfortunately, it is not uncommon for people to hold values, beliefs, and perceptions that are not conducive to safety, health, and environmental excellence. For example, the following negative beliefs are contrary to achieving SHE excellence and a systematic and supportive effort should be made to introduce people to a more positive way of thinking about SHE:

- Negative Belief: "Accidents and incidents are unavoidable."
  - Positive Alternative: "All accidents and incidents can be prevented through proactive and effective application of risk-management methodologies."
- Negative Belief: "Accidents and incidents are the cost of doing business."
  - Positive Alternative: "Accidents and incidents are unacceptable; we must operate our business without injuring employees and harming the environment."
- Negative Belief: "Realizing the goal of zero accidents and incidents is impossible."
  - Positive Alternative: "Achieving zero is possible if each employee takes charge of his or her own safety and works safely one minute, one day at a time."
- Negative Belief: "SHE is management's responsibility."
  - Positive Alternative: "Safety, health, and environmental protection are everyone's responsibility. We all play a key role in our own safety and the safety of others."
- Negative Belief: "SHE is not my job."
  - Positive Alternative: "SHE is everyone's job. SHE is a team game in which everyone plays a key role in achieving success. We can achieve more by working together as a team."
- Negative Belief: "We don't have time for safety."
  - Positive Alternative: "We don't have time for accidents and incidents."
- Negative Belief: "We don't have time to train our people."
  - Positive Alternative: "We can't afford not to train our people."

- Negative Belief: "We haven't had an accident yet so it must be safe."
  - Positive Alternative: "Just because we haven't had an accident doesn't mean there isn't a safer way to do the task."
- Negative Belief: "We've always done it this way."
  - Positive Alternative: "We are open to exploring new ways to make our workplace and work practices safer and more environmentally sound."
- Negative Belief: "It's only a minor accident."
  - Positive Alternative: "The severity of an accident is often due to chance; all accidents must be investigated so we can identify the root causes and prevent recurrence and the possibility of more severe accidents."
- Negative Belief: "It's only a small spill."
  - Positive Alternative: "There are no small spills, we must learn from all incidents and take steps to prevent them."
- Negative Belief: "Accidents and incidents are the employees' fault."
  - Positive Alternative: "Nothing is gained by placing blame. If employees have taken a risk or made an error, we must identify why and improve our management system so that the desired behavior is automatic."
- Negative Belief: "Safety is just common sense."
  - Positive Alternative: "Safety is not common sense; it is the careful and proactive application of risk management principles."
- Negative Belief: "Safety is a priority."
  - Positive Alternative: "Safety is an uncompromising organizational value."

In order to assess the state of an organization's SHE climate and the expressed attitudes and perceptions toward SHE in the organization, and to identify negative beliefs that are harmful and counterproductive to SHE excellence, it is often helpful to administer a formal SHE climate survey. According to Douglas Wiegmann et al.,[5] safety climate is "the temporal state measure of safety culture, subject to commonalities among individual perceptions of the organization. It is therefore situationally based, refers to the perceived state of safety at a particular place at a particular time, is relatively unstable, and subject to change depending on the features of the current environment or prevailing conditions."

It is relevant to assess safety climate because theoretical models and empirical evidence suggest that several organizational climate factors including management commitment, supervisor competence, priority of safety over production, time pressure, safety goals and standards, and participation factors affect safety performance.[6,7] In fact, Dan Petersen has argued that inasmuch as employee perceptions are a reality for an organization, a perception survey is "an invaluable diagnostic tool" that provides a stronger prediction of a company's future safety performance than injury rates or audits.[8,9] Perception surveys provide a means to measure leading indicators of SHE performance and provide a snapshot of the organization's current safety climate. A SHE perception survey enables one to determine what components of an existing safety, health, and environmental process are working well and what elements aren't, thus the organization is better equipped to identify what actions to take as it implements a Lean SHE initiative. A SHE perception survey also provides a baseline measurement of the organizational safety climate, which provides a starting point for measuring the progress of the SHE continuous improvement effort and culture change process. Today there are many commercially available perception surveys that have been validated as tools for assessing organizational safety climate.

If a formal SHE climate assessment is undertaken prior to launching the Lean SHE initiative, it is crucial that management be open to the input that employees provide. Ideally, the feedback should be viewed as a gift of insight provided by coworkers that can serve to make the organization better. Management should share and publicize the survey results via group meetings, sitewide postings, and visual charting of the findings. In no case should management keep the survey data secret or fail to act upon the findings. Leadership must be committed to acting upon the survey and should involve employees in the development of SHE action plans and problem solving based upon the climate assessment. If the assessment reveals specific gaps in leadership, employee beliefs, attitudes, practices, or SHE management systems, the design and implementation of the Lean SHE continuous improvement process should take these deficiencies into account. With new insights into the company's SHE climate and culture, management and coworkers are better prepared to design and implement an improved SHE process to achieve their shared vision of SHE excellence.

3.3 SHE Vision and Values

Our values are key drivers of our actions. Unlike our priorities, which are constantly changing depending upon the circumstances, our values are steadfast and deeply held. "Values are the anchors we use to make decisions so we can weather a storm. They keep us aligned with our authentic self. They keep us true to ourselves and the future we want to experience."[10] When people act based upon their values, there is more ownership of their actions and more commitment to the outcome. In fact, there is ample research showing that adaptable and values-driven companies are the most successful organizations.[11,12] Value-based leadership is concerned with:

1. How we apply our values in what we do
2. How we build on the values we share with others
3. How we create or add value has direct application in safety, health, and environment

As an organization embarks upon a Lean SHE improvement initiative it is important that the SHE vision and values of management and coworkers are fully aligned. It is crucial to identify and cultivate the organizational values that support SHE excellence. If different parts of the organization have divergent visions of SHE excellence and hold dissimilar values, it is difficult to achieve SHE excellence. Organizational discord of this nature is like a rowing team trying to win a race when everyone is paddling in a different direction; the chances of success are very slim. As previously discussed, a SHE climate assessment can provide insight into the SHE perceptions, beliefs, and values of people in the organization. If the climate analysis reveals that the SHE beliefs and values held by employees and management are divergent, or these beliefs and values are inconsistent with safety, health, and environmental excellence, efforts should be made to address this organizational schism.

Empowering a multidisciplinary employee vision and values team, consisting of both management and operating personnel, to develop and communicate a new organizational code that is supportive of SHE success is an important initial step toward realizing the vision of zero accidents and zero incidents. Jim Stewart has outlined a process for conducting future state visioning with the following steps:[13,14]

1. Create a list of key organizational stakeholders.
2. Develop a detailed description of the likely future environment.
3. Create a comprehensive future vision of what could be.
4. Contrast the future vision with the present state.
5. Express the values that will guide the organization as it seeks to realize its vision.
6. Express the vision in actionable terms.

Developing a shared SHE vision and values statement engenders employee ownership and commitment to SHE, and increases internal motivation to work in a safe and environmentally sound fashion. Ideally, when management and employees work together to develop a SHE vision and values statement, the document becomes more than trite words on a piece of paper, but rather a sincere expression of shared beliefs and values that compels everyone to action toward achieving SHE excellence. An example of a SHE vision and values statement is provided in Figure 3.3.

3.4 Zero Accident and Incident Education

Because it is common for many employees to believe that the goal of zero accidents and zero incidents is unattainable, it is useful during the preparation phase to conduct organization-wide training on achieving "zero." Zero accident and zero incident education focuses on

1. The organizational definition of zero accidents and incidents
2. Why achieving "zero" is the right SHE target
3. Why achieving "zero" is possible
4. Examples of organizations that have achieved "zero"
5. Our plan for achieving zero accidents and incidents

Organizations often state that their goal is zero accidents, but in reality, it is frequently just a lofty, politically correct platitude. No one in the organization believes that "zero" is a real target that brings any official accountability. In addition, there are often numerous differing definitions and understandings of what is meant by zero accidents or incidents. Does the "zero goal" refer to only no lost time accidents? Does it mean zero recordable accidents, or the loftier target of zero first-aid accidents? In light of this potential confusion it is important for organizational leadership to clearly define what is meant by the vision of zero accidents and incidents.

Many leading organizations define "zero accidents" to mean no recordable accidents over the course of a year because this target is SMART: specific, measurable, achievable, realistic, and time based. Because the definition of a recordable accident is clearly defined by OSHA and the Bureau of Labor Statistics (BLS) and their tracking is a regulatory requirement, this performance measure is generally reliable and comparable over time.

A Declaration of SHE Interdependence

Safety, health, and environmental excellence (SHE) is a crucial part of our organizational Vision and Values.

Our standard of SHE excellence, the vision to which we aspire as individuals and as an organization, is ZERO accidents and ZERO environmental incidents. NO accidents or incidents are acceptable, and ALL injuries can be prevented.

We believe that it is possible to work safely every day — without injuries, illnesses, or environmental incidents. The realization of this vision requires the establishment of a total safety culture in which ALL employees recognize SHE excellence as an unalterable value that is integral to our definition of world-class manufacturing and distribution. Focusing on SHE excellence throughout the organization is critical to delivering the vision that we have established.

Consistent with this vision the key articles or values of our Declaration of SHE Interdependence are

---

Value 1: People
People are our most important asset. Effective SHE systems focus on people.

Value 2: Interdependence
SHE excellence involves Interdependence.

Value 3: Zero is the Goal
Accidents and incidents are avoidable, zero accidents and incidents is achievable.

Value 4: Individual Responsibility
SHE is everyone's responsibility.

Value 5: Commitment
Commitment to safety, health, and environment is a job requirement.

Value 6: Teamwork
SHE excellence takes teamwork.

Value 7: Continual Improvement
SHE excellence requires a commitment to continuous improvement.

Value 8: Sustainability
We shall conduct our operations in a sustainable manner that meets the needs of the present, without compromising the needs of future generations.

Value 9: SHE is Good Business
SHE excellence provides competitive advantage and the ability to do good while doing well.

Figure 3.3   Example of a SHE vision and values statement.

The use of lost time accidents (LTAs) involving days away from work as a key performance indicator (KPI) for measuring safety success, however, is usually considered inappropriate. Lost time accidents are serious events involving significant injury; however, they do not measure the health of a safety process and are contrary to high levels of safety performance. Leading safety organizations should eliminate all serious accidents, thus the use of zero LTAs as the definition of "zero accidents" is not sufficiently rigorous and therefore is unsuitable as a measure of safety excellence. The use of LTAs as a measure of safety success also suffers from the temptation it provides to some organizations to engage in questionable case management practices in an attempt to avoid any lost time. The organizational focus is inappropriately placed on eliminating all lost time rather than eliminating all accidents. In extreme cases, ill-advised efforts can be made to manipulate and coerce injured employees to return to work against medical advice.

The use of no first-aid and no recordable accidents as the official KPI for measuring "zero accidents" also presents some concerns. If zero first-aid accidents are part of the organization's definition of safety success, there is the potential to discourage accident reporting, particularly at the start of the Lean SHE process. If first-aid accidents are not reported, the organization loses the opportunity to learn from these events and to prevent recurrence of the accident and eliminate more serious future accidents.

In the environmental area the definition of "zero" can take many forms. In the beginning stages of a "Lean and Green process" the definition of zero environmental incidents may mean no unpermitted environmental spills, discharges, or releases. As the organization's Green initiative matures and advances, the definition of "environmental zero" will expand to mean that no discharges and no waste of any kind, whether they are permitted or not, are considered acceptable. Progress is made in the environmental area by making incremental improvements in one operation at a time in order to convert each process to a zero-waste process. Once an organization adopts a zero-waste philosophy, dramatic changes in the way people think about waste and the environment occur. The old ways of doing business, where indiscriminant resource use and waste generation were thought of as the normal course of doing business, are no longer tolerable. With a new environmental ethic and zero-waste paradigm, the organization endeavors to convert the output or waste of one process into the input of another process. The zero-waste approach goes beyond basic recycling; it involves the redesign of products and processes so that individual product components and process outputs can be beneficially used in other processes. The zero-waste approach "imagines a future where everything is a renewable resource."[15] Zero-waste organizations passionately pursue the 3 Rs: reduce, reuse, and recycle. Zero waste strategies involve evaluating the entire life cycle of a product and the value stream of a process to identify improvement opportunities in the areas of

- Zero environmental incidents
- Zero solid and hazardous waste
- Zero emissions
- Zero effluent
- Zero waste of resources
- Zero toxics
- Zero energy wasted (100% energy efficiency)

3.5 SHE Loss Tree

You can't manage what you can't measure. Because this common management adage applies to the safety, health, and environmental area, it is important to develop a system for measuring and tracking SHE losses at the beginning of a SHE improvement initiative. Therefore, the development of a SHE loss tree is a key activity during the preparation phase of the Lean SHE Process. All organizations have limited resources; therefore, all SHE activities should be based upon SHE losses. The SHE management system and SHE work should focus on activities that reduce accidents, environmental incidents, and other SHE losses.

In TPM there are 16 major production losses, broken down as follows:[16]

A. Eight major equipment losses
   1. Breakdown or equipment failure loss
   2. Set-up and adjustment loss
   3. Cutting tool loss
   4. Start-up losses
   5. Minor stops and idling loss
   6. Speed losses
   7. Quality defect and rework loss
   8. Shutdown loss

B. Five big labor or human work efficiency losses
   9. Management loss
  10. Motion losses
  11. Line organization losses
  12. Logistics loss
  13. Measuring and adjustment loss
C. Three effective consumption effectiveness losses
  14. Yield losses
  15. Energy losses
  16. Die, jig, and tool losses

For these 16 major losses, a loss tree or branching diagram is commonly constructed to show graphically how the losses affect overall equipment efficiency (OEE), labor effectiveness, or resource consumption effectiveness.

Although accidents and environmental incidents are not specifically listed as one of the 16 major production losses, they can certainly be a cause of production losses. For example, a forklift accident in which the vehicle damages a machine can result in equipment breakdown loss. Accidents can cause both major and minor line stoppages, and can result in start-up losses and speed losses when less-experienced operators must replace injured workers. Under the traditional TPM loss tree, workplace accidents are classified as losses in addition to the 16 major losses. Accidents are considered a type of labor loss when employees cannot perform their full job tasks. Environmental losses such as waste disposal, spills, and energy losses are considered resource consumption losses.

Workplace accidents and environmental incidents are losses; therefore, the Lean and TPM philosophy requires that they be measured and tracked. As shown in Chapter 2 (Figure 2.4), a well-developed SHE loss tree enables an organization to track both the number and cost of each type of safety, health, and environmental loss. The safety losses that should be tracked include both the number and cost of all accidents of all severities, including

1. Fatalities
2. Lost-time accidents
3. Restricted work accidents
4. First-aid cases
5. Property-damage-only accidents

In addition to these safety loss events, the proactive tracking of pre-loss indicators such as near misses, at-risk behavior, and unsafe conditions is encouraged.

In the occupational health area, work-related illnesses of all severities should be tracked. In addition, proactive indicators of potential occupational illnesses should be monitored including exposures to

1. Chemicals
2. Ergonomic risks
3. Environmental conditions: heat, cold
4. Noise
5. Radiation

In the environmental area, environmental incidents of all types should be included in the SHE loss tree. In addition, discharges to all environmental media (air, water, land) should be quantified so that the organization's environmental footprint can be measured. The use of all types of energy and natural resources should also be measured.

Losses can be chronic and sporadic. As shown in Figure 3.4, chronic safety losses are generally long term, multicausal, and difficult to solve. In a safety process that is in control, the chronic safety losses represent the normal variation in the number of accidents that occur from month to month. The safety performance of the organization tends to vary between an upper control limit (UCL) and a lower control limit (LCL). According to Juran and Nakajima[17] the chronic losses of a process are a reflection of the underlying inefficiencies of the process. Thus, chronic accidents are a reflection of the inefficiencies inherent in the organization's safety management system. Chronic safety losses are the result of common causes, rather than special causes, within the management system.

Typically, innovative and fundamental changes in the organization's safety management system are required to break through this performance plateau caused by chronic safety issues. If an organization continues to utilize the same safety management system and to implement the same safety activities, safety performance will not improve significantly. Via the implementation of formal safety *kaizens* or improvements, a fundamental change in the safety process is made so that a significant drop in safety losses is achieved. This stepwise improvement in the organization's safety management system results in less variation in safety performance, and new and improved upper and lower control limits for accidents. Sporadic

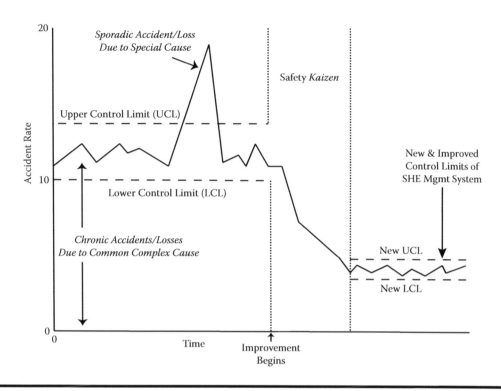

Figure 3.4   The use of focused improvement to address chronic safety losses. (Adapted from S. Nakajima, *TPM Development Program: Implementing Total Productive Maintenance.* **Productivity Press. Cambridge, MA, p. 39. With permission.)**

losses, or sudden peaks in accidents, are commonly caused by special causes or sudden changes in workplace conditions or systems. The key to addressing sporadic safety losses is to identify the workplace change and its associated root cause, and then to work to restore the workplace conditions and systems.

3.6 SHE Governance: Form Follows Function

Charles Darwin's theory of evolution introduced the concept that form follows function in biological systems. In other words, the anatomy of an animal reflects its intended function or use. For example, the long snout of the anteater reflects the need of the animal to reach insects underground and in hard-to-access places. The principle of form follows function is also applied to modern architecture and industrial design, which states that the shape of a building or object should be based upon its intended purpose. Renowned

American architect Louis Sullivan is credited with popularizing this aphorism in the following poem:

> *It is the pervading law of all things organic and inorganic,*
> *Of all things physical and metaphysical,*
> *Of all things human and all things super-human,*
> *Of all true manifestations of the head,*
> *Of the heart, of the soul,*
> *That the life is recognizable in its expression,*
> *That form ever follows function. This is the law.*[18]

The law of form follows function also applies to organizations. An organization that wants to operate quickly must be built for speed. Likewise, for an organization to function effectively in safety, health, and environment, its form needs to reflect this. The organization that aims for SHE excellence must have a structure and management system that is consistent with this goal. It is unlikely that a company will achieve SHE excellence if it devotes few resources to the area, has unclear SHE roles and responsibilities, and establishes no accountability. SHE governance refers to the structure and associated higher-level processes by which leaders are held accountable for SHE performance and through which the broadest strategic SHE decisions are made and implemented.

Figure 3.5 provides a SHE governance model that is appropriate for a Lean organization. SHE responsibility is embraced by leadership that serves on a SHE Council that establishes the organizational SHE vision, mission, and strategy. Through an interlocking team structure the SHE responsibility is cascaded through the line organization and the SHE strategy is put into action. A SHE pillar team with the support of SHE professionals develops the tactics and plans for implementing the overall SHE strategy. Special teams, with senior management sponsorship and support, are established to address key SHE issues affecting the organization such as sustainability. Line management has responsibility for implementing SHE and is assisted in this role by area SHE teams, SHE *kaizen* teams, and employees. This approach to SHE governance is consistent with the Lean philosophy because it

1. Recognizes that leadership commitment and involvement is essential for SHE success
2. Involves senior management in developing the SHE vision and strategy

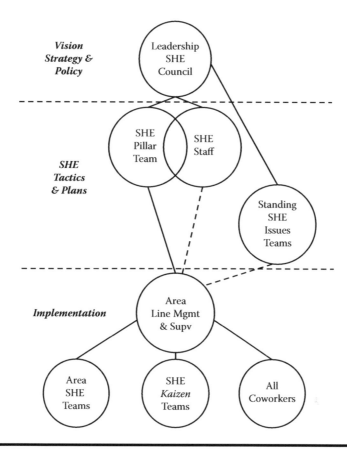

Figure 3.5   A SHE governance model.

3. Focuses resources on SHE issues that affect the organization and SHE losses
4. Uses an interlocking team structure to drive the SHE strategy and SHE improvement through the organization
5. Actively involves employees in the SHE process

3.7 SHE Targets and the Balanced Scorecard

Traditionally, SHE programs have utilized only lagging, performance-based measures, such as the number of accidents and incidents, to gauge success. Unfortunately, measurement and monitoring systems based solely upon this "body count and body part method" provide a limited view of the health of an organization's SHE process, and do not encourage the implementation of systematic risk-reduction programs and management systems. Accidents and incidents are lagging indicators that provide only a historical view of performance and do not provide the real-time information needed to manage risk. In organizations with outstanding SHE performance, accidents are rare events that provide a poor prediction of future performance. Management based solely upon accident and incident data is like trying to drive a car while looking in the rearview mirror. A view of the road you just traveled does not necessarily provide a good approach for navigating the road ahead. A better approach to assessing SHE success is to utilize both lagging accident and incident data and leading indicators to measure performance. Leading indicators of SHE performance include measures such as

1. The number of at-risk or unsafe behaviors observed
2. The number of unsafe conditions identified
3. The number of employees removed from risk
4. The number of environmental measurements above a designated action level

Success in safety, health, and the environment is not achieved by simple exhortations to do better, but rather the old-fashioned way: by doing work. The SHE work, however, cannot be any mundane activity or ritual—it must be work aimed at reducing risk and strengthening the organization's SHE management system and culture. Thus, the ideal SHE measurement and monitoring system should entail both performance- and activity-based measures that are aligned with the organization's overall strategy.

The use of the SHE balanced scorecard provides an ideal mechanism for establishing effective SHE targets because it facilitates the establishment of a performance measurement and monitoring system that

■ Includes both performance- and activity-based measures
■ Incorporates both leading and lagging indicators
■ Links SHE targets with the organization's strategy and vision

In business, a reliance on measurement systems that utilize only traditional, narrowly focused financial metrics can give a misleading view of performance, and can stifle improvement and innovation. Therefore, in many companies a balanced scorecard that incorporates both financial and nonfinancial operational metrics is utilized because it provides a more comprehensive view of business results and organizational performance. The balanced scorecard includes financial measures that summarize current business results, and operational metrics in areas that give an indication of organizational health and future performance. The business balanced scorecard provides management with a view of the organization from four perspectives: financial, customer, internal, and innovation and learning.[19,20]

Like the business balanced scorecard, the SHE balanced scorecard provides management with a more relevant, accurate, and comprehensive view of safety, health, and environmental performance by using a variety of metrics in five areas:

1. SHE performance results
2. SHE conditions (Machine)
3. SHE management systems (Method)
4. SHE behavioral measures (Man)
5. SHE culture (Milieu)

As shown in Figure 3.6, all SHE metrics in the balanced scorecard are aligned with the organization's SHE vision and strategy. The SHE balanced scorecard utilizes traditional safety results or outcome measures, but complements them with additional metrics that evaluate a different aspect of the

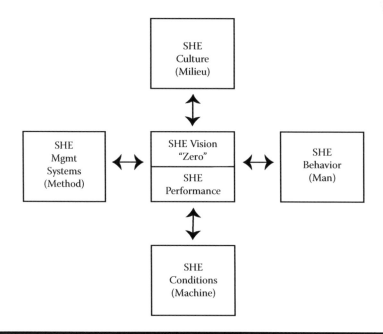

Figure 3.6   The SHE balanced scorecard model.

organization's SHE process. This approach provides leadership with a more comprehensive picture of (1) the state of the SHE system, (2) the SHE work that is necessary to achieve success, and (3) how to manage the SHE process.

It is a basic maxim of management that what gets measured gets done, and what gets measured and rewarded gets done well. The SHE balanced scorecard enables an organization to systematically determine what SHE work needs to get done, and how to measure this important work. Table 3.1 provides an example of how SHE measures, targets, and actions are linked via the SHE balanced scorecard, leading to the development of SMART (specific, measurable, achievable, realistic, timely) targets that are achieved through specific actions or work.

3.8 6S (5S for SHE)

Traditional workplace housekeeping and inspection programs often have limited success because they don't systematically identify the root causes behind substandard workplace conditions. A standard housekeeping program typically involves a few selected employees in conducting audits and generating lists of unacceptable items for someone else to fix. Unfortunately, this approach is reactive and does not foster ownership of workplace conditions because it doesn't involve all employees in the process of developing proactive housekeeping solutions. In addition, no attempt is made to identify and address the root causes of the substandard conditions, and so a vicious cycle is created where the same deficient conditions are repeatedly identified and corrected but never eliminated.

The basic 5S process is an introductory Lean methodology for creating and maintaining a clean, organized, and efficient workplace. The term 5S is derived from the five Japanese words and their English equivalents that describe the five basic steps of the process:[21,22]

Step 1: *Seiri* or Sort refers to putting the workplace in order by sorting needed items from unneeded ones. Via the technique of red tagging, area personnel inspect the workplace and tag items that are not needed. Work area clutter is eliminated and initial organization is established by removing unneeded items and storing those that are infrequently used.

Step 2: *Seiton* or Set means creating workplace arrangement and order by setting each work tool and item in its proper location. Establishing

Table 3.1 Model of a SHE Balanced Scorecard and Action Plan

No.	Objective Area	Measure	Target	Actions/"Work"
Performance Targets				
1	Safety	1.1 Recordable Injury Rate	50% reduction by year end vs. last year	1.1.1 Complete task-based risk assessments 1.1.2 Implement LOTO/ZES visual SOPs and OPLs 1.1.3 Error-proof chemical unloading operations
2	Health	2.1 Recordable Illness Rate	50% reduction by year end vs. last year	2.1.1 Complete ergonomic job evaluations 2.1.1 Complete noise mapping
3	Environment	3.1 Reportable Spills and Releases	50% reduction by year end vs. last year	3.1.1 Complete RCA of all previous spills, develop OPLs, and improvememnt plan
Management Systems Targets (Method/Management System)				
4	Safety	4.1 SHE Mgmt System Audit Score 4.2. SHE F-Tag Completion	Improve audit score 25% by year end 100% completion of all SHE F-tags within 1 week and category "A" SHE tags within 24 hours by mid-year	4.1.1 Develop and implement plan to address gaps identified in previous audit in risk assessment area 4.2.1 Produce daily Pareto chart of SHE F-tags and % completion chart

(Continued)

Table 3.1 Model of a SHE Balanced Scorecard and Action Plan (Continued)

No.	Objective Area	Measure	Target	Actions/"Work"
Management Systems Targets (Method/Management System)				
5	Health	5.1 Ergonomic (Ergo) Job Evaluations	Complete ergo evaluations for high-risk jobs by mid-year	5.1.1 Establish ergo teams to complete job evaluations
6	Environment	6.1 ISO 14001 Audit	Successful ISO recertification by May	6.1.1 Conduct internal ISO 14001 assessment 6 months prior to audit and develop plan to address gaps
Behavior-Based Targets (Man)				
7	Safety	7.1 At-Risk Observations	25% reduction by mid-year	7.1.1 Implement people-centered safety BBS process 7.1.2 Implement 1 BBS *kaizens* per month
8	Health	8.1 Ergo Observations	Develop ergo observation–feedback process in Q1	8.1.1 Identify key ergo behaviors, develop ergo observational definitions and observation sheet, and train observers
9	Environment	9.1 Environmental (Enviro) Observations	Initiate E observations in all areas by March 1	9.1.1 Identify key enviro behaviors, develop enviro observational definitions and observation sheet, and train observers

Equipment-/Condition-Based Targets (Machine)

10	Safety	10.1 Safety F-Tag *Kaizens*	Each area to complete 1 safety F-tag *kaizen*/month	10.1.1 Establish area safety *kaizen* teams 10.1.2 Implement 6S process
11	Health	11.1 Area Noise Levels	Reduce by 3 dB(A) by year end	11.1.1 Conduct area noise mapping 11.1.2 Identify high noise equipment 11.1.3 Implement noise *kaizens*
12	Environment	12.1 Carbon Footprint	Reduce by 10% per year	12.1.1 Establish carbon baseline 12.1.2 Identify equip. with largest carbon footprint 12.1.3 Implement carbon/energy *kaizens*

Cultural Targets

13	Safety	13.1 Safety Climate Survey	Establish SHE climate baseline by Q2	13.1.1 Conduct formal safety climate survey and summarize results
14	Health	14.1 Employee Absentee Rate	Reduce 10% by year end	14.1.1 Implement smoking cessation program 14.1.2 Conduct monthly wellness activities 14.1.3 Provide wellness counseling
15	Environment	15.1 Recycling	Increase 25% by year end	15.1.1 Establish employee Green teams to promote recycling and conduct enviro *kaizens*

a place for everything and having everything in its place is the goal of this step. Storage locations and limits are established and clearly marked in a visual fashion. Frequently used items are clearly visible and readily accessible.

Step 3: *Seiso* or Shine involves thoroughly cleaning the workplace so that everything shines. Via this step workers develop pride in and ownership of their work area.

Step 4: *Seiketsu* or Standardize comprises the establishment of written workplace standards for cleanliness and orderliness that codify the agreed procedures and best practice. In addition, visual controls are put in place to facilitate compliance with 5S standards.

Step 5: *Shitsuke* or Sustain entails developing the organizational discipline and processes to sustain the high 5S standards for workplace cleanliness and orderliness. 5S is maintained via training, education, empowerment, enforcement, promotion, and commitment. 5S maps, checklists, and audits are used to sustain high levels of 5S.

Through the rigorous application of basic 5S, workplace orderliness and cleanliness are dramatically improved. Sites that strive for a high standard of 5S adopt visual workplace methods that apply visual and standardized approaches to the marking and storage of all equipment. When sites initially implement a 5S program they naturally tend to focus on improvements to general plant and equipment conditions, rather than safety issues. However, after a site becomes more experienced in basic 5S techniques there is value in expanding the 5S focus into the SHE area.

The five steps of the basic 5S process should also be applied to the storage and organization of facility safety, health, and environmental equipment. Table 3.2 provides examples of how 5S visual methods can be applied to the storage and organization of SHE equipment.

Although the basic 5S process does not focus directly on safety, health, and environment, by improving overall workplace housekeeping and orderliness, it tends to indirectly improve site SHE conditions. Many organizations have modified the basic 5S program by adding a sixth S (safety, health, and environment) so that the 5S principles and methodology can be leveraged to improve safety, health, and environmental conditions and ultimately SHE performance. This expanded 5S process, commonly referred to as 6S or "5S for SHE,"[23] is a formal process that provides specific techniques and tools for expanding a 5S program into the SHE area.

Table 3.2   Application of Visual 5S Methods to Safety, Health, and Environment

Equipment Type	5S Visual Storage/Marking Method
Safety Equipment	
Fire Extinguishers and Equipment	Post signs and paint red boxes under the equipment.
Safety Showers/Eye Wash	Post signs and paint green boxes on the floor.
Emergency Respirators	Use of yellow storage cabinets.
Personal Protective Equipment	Use of PPE storage cabinets with shadow boards.
Lockout/Tagout Devices	Hang LOTO devices on peg shadow board.
Right-to-Know Stations	Post MSDS sign; paint green and white lines on the wall.
Safety Walkways	Paint yellow or green lines on the floor marking walkway.
Health Equipment	
AEDs	Post sign and store in a standardized red cabinet.
First-Aid Supplies	Post sign and store in inventoried white cabinet with red cross.
Environmental Equipment	
Spill Supplies	Post sign and place in inventoried blue container.
Sanitary Drain	Drain covers are painted yellow and stenciled sanitary drain.
Storm Drains	Drain covers are painted green and stenciled storm drain.

The purpose of 6S is the systematic improvement of workplace safety, health, and environmental conditions. Trained area personnel develop a 6S implementation plan for a targeted area of the workplace. The team inspects the work area with a focus on SHE physical conditions rather than SHE behaviors. Whenever a substandard safety, health, or environmental condition is identified, a uniquely colored, multipart safety tag (often green) is completed. In many organizations the tag is referred to as a safety F-tag or safety Fault-tag. A copy of the tag is applied to the equipment in order to notify site personnel of the hazard. Another copy of the tag is sent to site maintenance so that the substandard condition can be scheduled for repair, and a third copy of the tag is retained by the area 6S inspection team.

The 6S team places a copy of the tag on the area 6S activity board and records the location of the at-risk condition on a map of the work area. As shown in Figure 3.7, the team creates a dot or "measles" map marking the location of all the substandard SHE conditions identified during the audit. The dot map provides a useful baseline picture of the state of SHE in the workplace. In addition, the visual indication of the distribution of substandard conditions is helpful in determining if there is one particular area or piece of equipment associated with a high number of SHE hazards. This information can be helpful in identifying the source or root cause of SHE issues.

A classic example of how a dot map can be used to identify the source of a safety, health, or environmental hazard is the study of cholera cases in London by Dr. John Snow in 1854.[24] By creating a dot map that showed the location of all cholera cases in the Soho section of the city, Dr. Snow

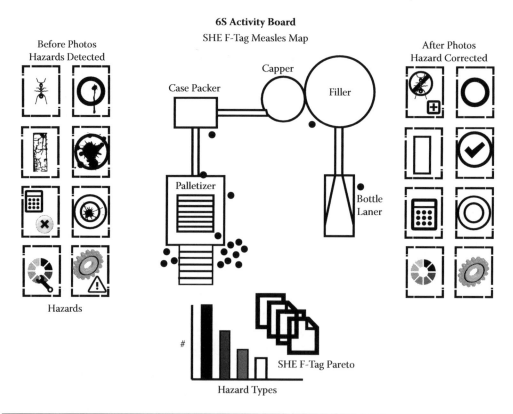

Figure 3.7 Model of a 6S activity board with measles map.

was able to pinpoint the source of the outbreak. Despite the general lack of acceptance and understanding of the germ theory of disease at the time, Dr. Snow's visual dot map was able to convince authorities that the source of the cholera outbreak was the public water pump on Broad Street because the cholera cases were clustered in the neighborhood surrounding this pump. Dr. Snow's study of the distribution of the disease was convincing enough to persuade local officials to eliminate the source of the contaminated water by removing the pump handle and thereby disabling the well. In a similar, but admittedly less dramatic fashion, a dot map of at-risk conditions identified during 6S audits can provide useful clues to the source of hazards.

After constructing the dot map the area team studies the distribution of substandard conditions to identify the equipment or areas that may be the source of problems. In addition, the team categorizes the at-risk conditions and performs a Pareto analysis to identify the most common hazards. Formal root cause analysis (RCA) is conducted to identify the root causes of the at-risk conditions. Once the root causes are identified, the team conducts formal problem-solving or *kaizen* activities in order to implement corrective and preventive measures that will eliminate the substandard conditions. Key learning and leading practices are documented in standards, communicated, and replicated throughout the organization. In this way, the 6S process serves to systematically identify and remove at-risk conditions from the workplace.

In summary, the steps of the 6S (or 5S for SHE) process involve the following activities:

1. The area team receives initial training on the 6S methodology.
2. An area 6S implementation plan is developed.
3. The area team conducts a focused 6S audit in order to systematically identify substandard workplace SHE conditions.
4. Multicopy SHE tags are completed for each at-risk condition identified.
   One copy is used to mark the location of the hazard.
   One copy is forwarded to the maintenance department.
   One copy is placed on the area 6S activity board.
5. The team posts a dot or "measles" map on the area 6S Activity board showing the location of all at-risk conditions.
6. The area team conducts a Pareto analysis of all SHE tag at-risk conditions and posts a chart of common substandard conditions.

7. The team conducts a root cause analysis of the most common at-risk conditions.
8. The team conducts *kaizen* (improvement activity) to address the root causes of at-risk conditions.
9. Standards are developed to document leading practices.
10. The team communicates its findings and key learning via meetings and OPLs.

3.9 Lockout/Tagout and Zero Energy State (LOTO and ZES)

Lean and TPM activities including cleaning, inspecting, adjusting, and maintenance require facility personnel to come into close contact with plant equipment. Typically this equipment contains many hazardous energy sources such as electrical, mechanical, pneumatic, hydraulic, chemical, and gravitational energy. In light of this, it is necessary during the preparation phase of Lean to ensure that site personnel know how to properly de-energize and secure all machinery with which they will be working. The process of shutting off equipment, de-energizing it, placing it in a safe state, and securing equipment so that it cannot be turned on is known as lockout/tagout and zero energy state.

A key philosophy of Lean and TPM is that safety comes first. Therefore, before any person works on equipment, knowledgeable employees should develop visual LOTO and ZES procedures and one-point lessons (OPLs). As shown in Figure 3.8, unique symbols are used for each energy source. The visual LOTO and ZES procedures show the location of each energy shutoff device by placing the symbol on equipment diagrams or digital photographs of the equipment. In this way, detailed written safety procedures are simplified via the use of visual symbols and diagrams.

In some organizations 5S and LOTO/ZES is considered Step 0 of autonomous maintenance (AM), indicating that no AM activities can begin until these essential safety activities are performed. All employees who are to conduct AM activities must be trained in LOTO/ZES and must demonstrate that they can properly de-energize and secure the equipment with which they will work. Understanding how a machine is properly de-energized is the first step in gaining a deep understanding of equipment operation, and in taking ownership of the machine.

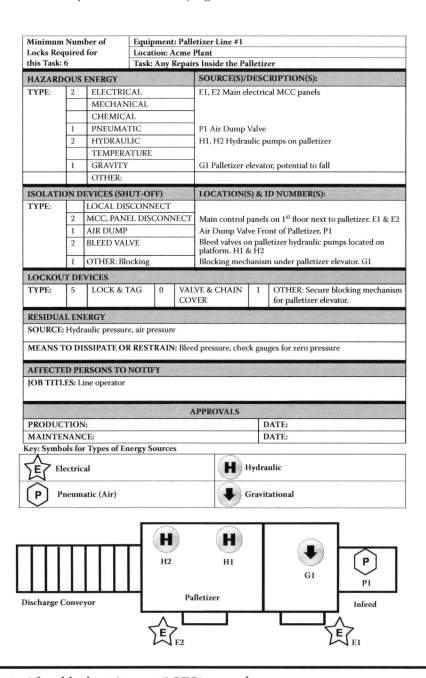

Minimum Number of Locks Required for this Task: 6	Equipment: Palletizer Line #1
	Location: Acme Plant
	Task: Any Repairs Inside the Palletizer

HAZARDOUS ENERGY			SOURCE(S)/DESCRIPTION(S):
TYPE:	2	ELECTRICAL	E1, E2 Main electrical MCC panels
		MECHANICAL	
		CHEMICAL	
	1	PNEUMATIC	P1 Air Dump Valve
	2	HYDRAULIC	H1, H2 Hydraulic pumps on palletizer
		TEMPERATURE	
	1	GRAVITY	G1 Palletizer elevator, potential to fall
		OTHER:	

ISOLATION DEVICES (SHUT-OFF)			LOCATION(S) & ID NUMBER(S):
TYPE:		LOCAL DISCONNECT	
	2	MCC, PANEL DISCONNECT	Main control panels on 1st floor next to palletizer. E1 & E2
	1	AIR DUMP	Air Dump Valve Front of Palletizer, P1
	2	BLEED VALVE	Bleed valves on palletizer hydraulic pumps located on platform. H1 & H2
	1	OTHER: Blocking	Blocking mechanism under palletizer elevator. G1

LOCKOUT DEVICES						
TYPE:	5	LOCK & TAG	0	VALVE & CHAIN COVER	1	OTHER: Secure blocking mechanism for palletizer elevator.

RESIDUAL ENERGY

SOURCE: Hydraulic pressure, air pressure

MEANS TO DISSIPATE OR RESTRAIN: Bleed pressure, check gauges for zero pressure

AFFECTED PERSONS TO NOTIFY

JOB TITLES: Line operator

APPROVALS		
PRODUCTION:		DATE:
MAINTENANCE:		DATE:

Key: Symbols for Types of Energy Sources

E	Electrical	H	Hydraulic
P	Pneumatic (Air)	⬇	Gravitational

Figure 3.8   Visual lockout/tagout (LOTO) procedure.

Endnotes

1. Sun Tzu (2005). *The Art of War*. English translation by Lionel Giles. El Paso Norte Press, El Paso, TX.
2. Plato (427–347 BC). *The Republic*. Penguin Classics, New York, Sept. 1955.
3. Schweitzer, Melodie (Nov. 2007). Creating a safety culture through felt leadership. *Industrial Hygiene News*.
4. Wattleton, Faye (2009). WhatQuote.com. Retrieved Sept. 1, 2009 from: http://www.whatquote.com/quotes/Faye-Wattleton/16448-The-only-safe-ship-i.htm
5. Wiegmann, Douglas A., Zhang, H., von Thaden, T., and A. Mitchell (2002). "A synthesis of safety culture and safety climate research." University of Illinois at Urbana-Champaign & Federal Aviation Administration Technical Report ARL-02-3/FAA-02-2. June 2002. Retrieved on Sept. 1, 2009 from: http://www.humanfactors.uiuc.edu/Reports&PapersPDFs/TechReport/02-03.pdf
6. Flin, R., Mearns, K., O'Conner, P., and R. Bryden (2000). Measuring safety climate: Identifying the common features. *Safety Science*, 34: 177–192.
7. Oliver, A., Tomás, J. M., and A. Cheyne (2006). Safety climate: Its nature and predictive power. *Psychology in Spain*, 10: 28–36.
8. Johnson, Dave (2003). Perception is reality. *Industrial Safety and Hygiene News ISHN E-News*, 2, 30, Friday, October 3, 2003.
9. Petersen, Dan (1988). *Safety Management: A Human Approach*. Aloray, Inc., Goshen, NY.
10. Barrett, Richard (2009). "Values based leadership: Why is it important for the future of your organization?" Barrett Values Centre. Retrieved on Sept. 1, 2009 from: www.valuescentre.com/docs/ValuesBasedLeadership.pdf
11. Majer, Kenneth (2004). *Values-Based Leadership: A Revolutionary Approach to Business Success and Personal Prosperity*. Majer Communications, San Diego.
12. Hess, Edward D. and Kim S. Cameron (2006). *Leading with Values: Positivity, Virtue and High Performance*. Cambridge University Press, Cambridge, UK.
13. Stewart, James M. (2002). *Managing for World Class Safety*. Wiley-Interscience, New York.
14. Stewart, James M. (1993). Future state visioning—a powerful leadership process. *Long Range Planning*, 26, 6: 89–98.
15. Winter, Jessica (2007). "A world without waste." *Boston Globe*. March 11, 2007. Retrieved on September 15, 2010 from: http://www.boston.com/news/education/higher/articles/2007/03/11/a_world_without_waste/
16. Shirosi, Kunio (1996). *Total productive maintenance: New implementation program in fabrication and assembly industries*. Japan Institute of Plant Maintenance (JIPM). pp. 40–70.
17. Nakajima, Seiichi (1989). *TPM Development Program: Implementing Total Productive Maintenance*. Productivity Press. Cambridge, MA, pp. 39–57.

18. Sullivan, Louis H. (1896). "The tall office building artistically considered." *Lippincott's Magazine*, March 1896, in MIT Open Courseware. Retrieved from http://ocw.mit.edu/OcwWeb/Civil-and-Environmental-Engineering/1-012Spring2002/Readings/detail/-The-Tall-Office-Building-Artistically-Considered-.htm
19. Kaplan, Robert and David Norton (January–February, 1992). The balanced scorecard—Measures that drive performance. *Harvard Business Review*, (pp. 71–79).
20. Kaplan, Robert and David Norton (December 23, 2002). Partnering and the balanced scorecard. Harvard Business School Working Knowledge. Retrieved from http://hbswk.hbs.edu/item/3231.html
21. Productivity Europe (1998). *The 5S Improvement Handbook*. Productivity Europe Ltd. Bedford, UK.
22. Graphic Products, Inc. (2009) "The 5S Philosophy." Beaverton, OR. Retrieved Sept.1, 2009 from www.GraphicProducts.com
23. Millard, Lorraine (1999). *5S for Safety: New Eyes for the Shop Floor*. Primedia Workplace Learning.
24. Johnson, Steven (2006). *The Ghost Map: The Story of London's Most Terrifying Epidemic—And How It Changed Science, Cities and the Modern World*. Riverhead Books/Penguin Group, New York.

Chapter 4

Integrating SHE into the Autonomous Maintenance Pillar

Better a thousand times careful than once dead.

Chinese proverb

4.1 Overview of the Integration of Autonomous Maintenance and SHE

Autonomous maintenance (AM), or *Jishu-hozen* in Japanese, is commonly referred to as the pace-setting pillar inasmuch as it plays a key role in establishing the speed of the Lean/TPM process implementation and the resulting rate of operational improvement and culture change. The term *maintenance* in autonomous maintenance does not to refer to craftpersons fixing equipment, but rather to operating personnel maintaining or keeping equipment in good running order. AM, a basic building block of a Lean/TPM process, consists of the following seven steps that aim to improve both plant equipment and plant personnel, and thereby increase overall equipment effectiveness (OEE):

Step 1: Initial clean
Step 2: Countermeasures to sources of contamination

Step 3: Tentative AM standards
Step 4: General inspections
Step 5: Autonomous inspection or checking
Step 6: Standardization
Step 7: All-out AM management

The steps of AM build upon each other and form a systematic process in which operators gain a deeper understanding of their equipment, assume ownership of their equipment, and take measures to ensure optimal equipment performance. A key outcome of AM is that the equipment operates better, and as a result the operator's job is made easier and safer because he or she does not have to adjust and fix the equipment as frequently. Thus, via simple implementation of AM activities, improvements in safety, health, and environmental performance are achieved.[1,2]

By systematically integrating the autonomous maintenance and the SHE pillar, an organization can achieve further improvements in overall equipment effectiveness and in SHE. Both autonomous maintenance and SHE achieve success by providing operators with the requisite knowledge, skills, and tools to take ownership of their equipment. Similar to autonomous maintenance, the concept of autonomous safety is the state in which operators have the skills, knowledge, and motivation to work safely. Via autonomous maintenance and autonomous safety, operators are empowered, take ownership of their equipment and its safety, and actively make both operational and safety improvements on the production line. In his book, *TPM in Process Industries*, Tokutaro Suzuki emphasizes that because it takes time to establish a SHE culture, it is important to incorporate safety, health, and environment into autonomous maintenance at the earliest stages. According to Suzuki, "Safety awareness takes time to sink in (therefore) it is important to begin addressing it from the preparation phase."[3] Therefore, during each AM step specific autonomous maintenance tools and methodologies should be utilized and leveraged to improve SHE performance. Figure 4.1 shows the different SHE activities that can be implemented during each successive AM step in order to drive SHE performance and incorporate SHE into the organizational culture.[4]

4.2 AM Step 1: Initial Cleaning and SHE

The first step of autonomous maintenance, initial cleaning, involves teams of operators in the thorough deep cleaning of their equipment where

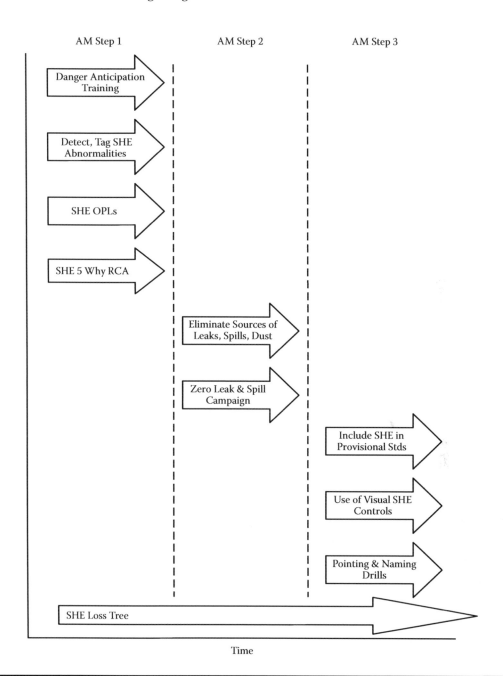

Figure 4.1 SHE activities undertaken during AM Steps 1–3.

every component of the machine is touched in order to expose equipment abnormalities.[5] During AM Step 1 operators develop and execute a detailed equipment cleaning plan, tag equipment defects that are identified, explore reasons behind the defects, and implement some initial improvements. As previously discussed, these activities bring workers into intimate contact with machinery and require them to place their body parts into the operating zone of equipment; therefore, it is essential that operators receive lockout/tagout and zero energy state (LOTO/ZES) training and basic safety instruction prior to beginning AM Step 1. In addition, the team leader should ensure that equipment and task-specific safety instructions are given to all operators before beginning each AM task. Many organizations also find it useful to have the operator team recite a daily group safety pledge that they have developed, such as "Safety comes first," to remind them of their team commitment to zero accidents and zero incidents.

4.2.1 Danger Anticipation and Experiential Training

In addition to LOTO/ZES training and general safety instructions, it is useful to provide operators with safety training that enables them to anticipate dangers or hazards specific to the equipment on which they will be working. The belief that training on danger anticipation will prevent accidents is based upon the safety brain theory developed by Mitsuo Nagamachi.[6] In his article, "Accident Prevention Management Based on Brain Theory," Nagamachi argues that the human brain is divided into two layers that have very different functions: an upper layer called the neo cortex (*Shiromiso* or white brain) and a central section called the old cortex (*Akamiso* or red brain). The old cortex, which is common among all species of animals, functions based upon instinct and emotion. On the contrary, the neo cortex, which exists only in the mammalian animal brain, is a center of recognition, memory, and judgment. It is believed that unsafe behavior stems from the emotional mind of the old cortex rather than the more analytical neo cortex. Due to differences in people's brains and which section is being utilized, the meaning of visual perception and the anticipation of danger differs from person to person. A person operating from the analytical neo cortex may perceive a situation as presenting risk, whereas a person operating from the emotional old cortex will fail to identify the danger. Thus, to a large extent, the perception of danger and risk is dependent upon a person's emotional attitude or state.

In light of his safety brain theory, Nagamachi has developed an approach to safety training that improves a person's ability to recognize danger. The training method involves showing workers pictures, videos, or simulations of dangerous situations related to their work and having them discuss what they see in the pictures. The theory is that workers' ability to recognize danger in their actual work situation is improved if they have seen a similar situation in training, because they are more likely to approach the situation using their analytical neo cortex rather than to act rashly using their emotional brain or old cortex. For example, pilots who are trained on a flight simulator that presents them with various emergency situations are more likely to act calmly and analytically when confronted with the real danger. In a similar fashion, by using photos or videos rather than expensive simulators, danger anticipation training can be provided to workers in any job. For instance, the ability of truck drivers to anticipate and successfully respond to danger can be improved by showing them images or videos of dangerous driving situations they are likely to encounter, and then facilitating safety discussions about these situations. When drivers are shown a photo or video of a ball bouncing into the street and are asked what potential danger exists, their ability to recognize the danger and respond to it improves. A discussion can then be facilitated about the danger of children running into the roadway and the signals that enable a driver to anticipate this danger. Likewise, video and images of vehicles rapidly approaching intersections can be used to improve driver recognition of intersection dangers. Carefully and safely staged video or pictures of what was happening just prior to a previous accident can also be used to enable workers to anticipate and respond correctly to dangerous situations.

Another useful SHE activity to improve danger anticipation is to have teams of operators walk through the *gemba* or workplace in order to identify and discuss potential safety, health, and environmental hazards associated with their equipment, particularly hidden hazards. A simple safety checklist can be used to ensure that the team identifies key hidden hazards. Afterward, the location of potential machine danger points can be made clearly visible via the use of safety labels. In addition, it is useful to brainstorm and discuss potential safety, health, and environmental issues that could arise as a result of AM Step 1 initial cleaning activities.

Unfortunately, some people believe that an accident will never happen to them, and therefore will resist altering their unsafe work practices. Often it takes a dramatic personal experience with a life-threatening accident to change their unsafe attitudes and behaviors. Too familiar is the story of a company or site with a poor safety record whose management and employees got "safety religion" and dramatically improved their safety performance only after experiencing a horrible, catastrophic, and life-altering accident. The emotional and tragic event left an indelible safety impression on the hearts and minds of all involved, making safety a personal and unalterable value for everyone in the organization.

Organizations, however, can ill afford to wait until all recalcitrant employees experience a life-altering accident and undergo a personal safety transformation. The challenge for organizations is to enable employees to have an emotional and safety-promoting experience without being involved in an accident. Experiential safety training aims to do this by providing participants with an emotional and thought-provoking training experience that permanently impresses upon them the importance of safe work practices. Some experiential training involves realistic role-playing exercises that enable an employee to personally experience a potential accident scenario and its tremendous cost to self and family. Modern virtual reality and three-dimensional (3D) technologies now provide the means to make safety training so realistic that it provokes an emotional and memorable response in the trainee that results in greater understanding of safety risks, increased safety commitment, and improved safety performance. Experiential training also improves danger anticipation because it provides the employee with the opportunity to safely experience what can go wrong during the performance of hazardous job tasks.

4.2.2 SHE F-Tags

AM Step 1 involves operators in "cleaning with meaning." In other words, equipment cleaning is not performed in a random and meaningless fashion, but rather with the purpose of uncovering hidden defects. In AM Step 1, cleaning encompasses inspection because it enables the operator to identify equipment abnormalities and their source. Thus, the operators identify latent equipment defects and take initial steps to address them. In addition, the operators come to better understand their machinery and how it works, and begin to take ownership of their equipment and responsibility for its safe operation.

The key activity of AM Step 1 is the deep cleaning of equipment in order to identify machine abnormalities such as leaks, contamination, loose parts, and damaged parts. When defects are identified they are labeled with brightly colored tags, commonly referred to as F-tags (*fuguai* or fault/abnormality tags) or M-tags (maintenance tags). Two types of F-tags of different colors are commonly used: operator F-tags and maintenance F-tags. Operator tags are used to identify equipment abnormalities that can be addressed by production workers. Maintenance tags are generally used to visually highlight equipment faults that require the intervention of a trained mechanic. In order to emphasize the identification of safety, health, and environmental hazards or abnormalities some organizations establish a third type of F-tag, a SHE F-tag. Figure 4.2 provides an example of a SHE F-tag and a one-point lesson on completing this type of F-tag.[7]

Safety, health, and environmental defects are identified by deep cleaning of the equipment and its components and by using four of the senses: sight, hearing, touch, and smell. When a SHE defect is identified, a copy of the brightly colored SHE F-tag is placed on the equipment to visually mark the location of the SHE hazard or abnormality. The F-tags are logged and a visual SHE defect map marking the location of the hazards is developed and posted. The SHE F-tags are categorized and Pareto charts of the SHE issues identified are posted. "5 Why" root cause analyses (RCA) are performed by a team in order to identify the underlying causes of the defects. Tentative countermeasures are then implemented to address the SHE abnormalities.

4.2.2.1 One-Point Lessons (OPLs)

Learning from each SHE root cause analysis is communicated and transferred by means of one-point lessons (OPLs) or single-point lessons. An OPL is a one-page training document designed to teach a single concept or skill. Typically the OPL uses diagrams or pictures to visually demonstrate the lesson rather than long worded procedures. Figure 4.3, a safety OPL on how to make adjustments on grinding wheel guards, provides an example of how digital photographs can be used to visually and effectively convey a safety procedure. Once developed, the OPL is reviewed and discussed with each employee performing the task. In addition, it is helpful to have employees go to the *gemba* to observe and demonstrate the safety concept or skill. When employees demonstrate that they understand the safety concept of the OPL, they initial the document. As the AM process progresses and the team identifies key safety, health, and environmental issues there is significant value in having the team develop SHE OPLs that clearly and visually display the proper and safe way to perform critical tasks.[8]

4.2.3 Incorporating Safety into Cleaning and Inspection Plans

As part of AM Step 1, operators develop cleaning and inspection plans and checklists for their equipment in order to establish basic machine conditions and to expose and eliminate hidden defects. In an effort to integrate SHE into

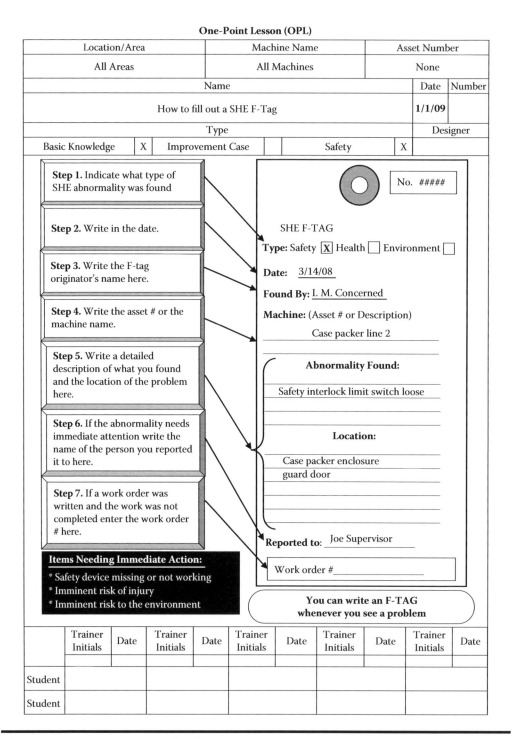

One-Point Lesson (OPL)

Location/Area	Machine Name	Asset Number
All Areas	All Machines	None

Name	Date	Number
How to fill out a SHE F-Tag	**1/1/09**	

Type				Designer
Basic Knowledge	X	Improvement Case		Safety	X	

Step 1. Indicate what type of SHE abnormality was found

Step 2. Write in the date.

Step 3. Write the F-tag originator's name here.

Step 4. Write the asset # or the machine name.

Step 5. Write a detailed description of what you found and the location of the problem here.

Step 6. If the abnormality needs immediate attention write the name of the person you reported it to here.

Step 7. If a work order was written and the work was not completed enter the work order # here.

Items Needing Immediate Action:
* Safety device missing or not working
* Imminent risk of injury
* Imminent risk to the environment

No. #####

SHE F-TAG

Type: Safety [X] Health □ Environment □

Date: 3/14/08

Found By: I. M. Concerned

Machine: (Asset # or Description)
Case packer line 2

Abnormality Found:
Safety interlock limit switch loose

Location:
Case packer enclosure
guard door

Reported to: Joe Supervisor

Work order #

You can write an F-TAG whenever you see a problem

	Trainer Initials	Date	Trainer Initials	Date	Trainer Initials	Date	Trainer Initials	Date	Trainer Initials	Date
Student										
Student										

Figure 4.2   SHE F-tag one-point lesson.

SAFETY ONE-POINT LESSON (OPL)				
Title: **Proper Grinder Gap Adjustments**			Date: 01/01/09	
Class: Specific Knowledge [X] Improvement [ ] Safety [X]			OPL# Shop #1	

Actual Results	Date																
	Student																

Parts of a Grinding Wheel:

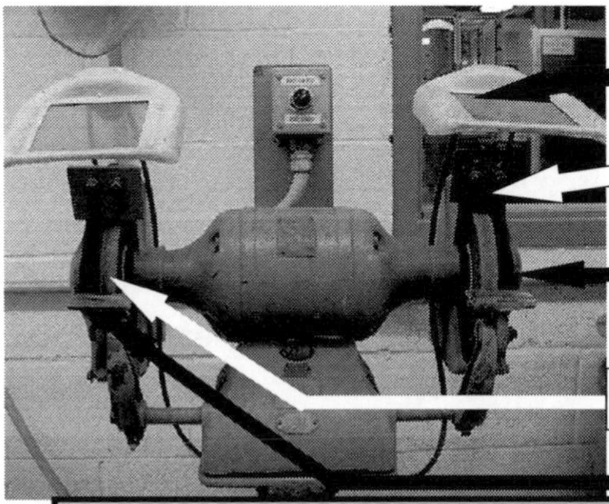

Grinder Shield

Tongue Guard

Grinding Wheel Cover

Grinding Wheel Face

Tool Rest

Safety Procedure: Before & after use check all grinder guards and adjust the tongue guard and tool rest to ensure that the gaps meet the standards specified.

PPE: Always use the grinder shield and a face shield when operating the grinder

Tongue guard adjustment: Loosen these two bolts here.

1/4 inch maximum gap

1/8 inch maximum gap

Tool rest adjustment: Loosen bolts here (lower bolt not shown).

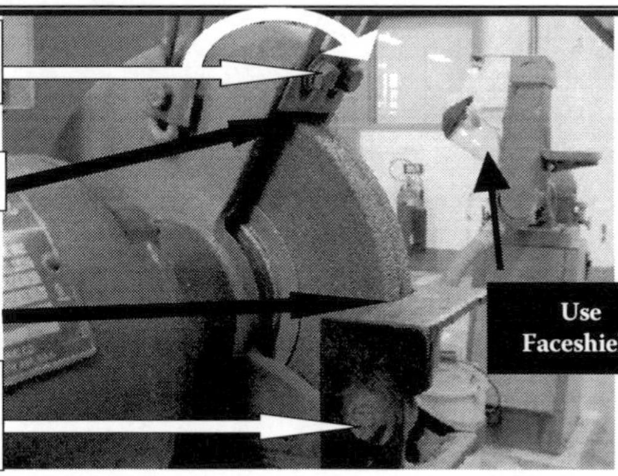

Use Faceshield

Figure 4.3 Example of SHE one-point lesson.

the TPM AM activities it is useful to include the checking and cleaning of key safety items on the cleaning plan. The checking and cleaning of equipment safety controls including machine interlocks and guarding should be included on the checklist. In addition, a list of potential safety abnormalities that should be checked on specific pieces of equipment can be developed by analyzing the safety F-tags for that equipment. In this way safety is not managed as a separate activity, but becomes an integral part of the autonomous maintenance process.

4.3 AM Step 2: Countermeasures to Sources of Contamination and SHE

AM Step 2 aims to reduce cleaning time and improve equipment conditions by implementing countermeasures to sources of contamination and hard-to-access areas. Sources of contamination such as dust, dirt, oil, and leaks are painstakingly rooted out and then eliminated or contained by means of temporary countermeasures or improvements. Areas that are difficult to clean, lubricate, and inspect are improved.

4.3.1 Elimination of Sources of Leaks, Spills, and Dust

The first goal of AM Step 2—to identify and address sources of contamination—has a direct impact on reducing equipment leaks, spills, and releases of all types of materials such as oil, grease, chemicals, product, dust, and dirt. The elimination of equipment leaks, spills, and releases has many obvious SHE benefits including

1. Improved safety as the result of the elimination of slip, trip, and fall hazards
2. Improved occupational health through the reduction in chemical exposure
3. Improved environmental performance due to fewer spills and releases into the environment

4.3.2 Elimination of Hard-to-Access Areas

The second goal of AM Step 2—improving access to areas that are hard to clean, lubricate, inspect, or tighten—also has clear safety, health, and environmental benefits. One important safety benefit is that potential employee strains are eliminated through more ergonomic arrangement of equipment.

Examples of countermeasures that eliminate hard-to-access equipment and improve safety include

1. Installation of automatic lubrication systems that eliminate the need for employees to navigate cramped and tight locations in order to manually lubricate equipment
2. Relocation of equipment gauges to eye level and outside the equipment to enable easy inspection and monitoring of machine functioning

4.3.3 Zero Leak and Spill Campaign

The tools and methodologies employed in AM Step 2 can be leveraged to improve environmental performance by launching a zero environmental leak and spill campaign. The goal of the campaign is to eliminate all environmental spills and leaks in a specific area of the facility. An improvement team should be formed and chartered with the goal of achieving zero leaks and spills. The improvement team should utilize a formal problem-solving methodology such as the PDCA (plan–do–check–act) cycle or the CAPDo (check–analyze–plan–do) process to systematically identify and remove the root causes of all spills and leaks.[9] An activity board is posted to communicate to the entire workforce the progress of the improvement project. The steps of a CAPDo zero leak and spill campaign are as follows:

1. CHECK: Conduct an area audit to identify all leaks and spills.
   a. Visually mark the location of all spills and leaks with a SHE F-tag, cone, or spill sign to inform all coworkers of potential slip, trip, and fall hazards.
   b. Place a completed SHE F-tag on the source of all identified spills and leaks.
   c. Create and post a measles map of the work area identifying the location of all spills and leaks.
2. ANALYZE: Categorize the types of spills and leaks, and complete a Pareto analysis of the types of spills or leaks.
   a. Develop a ranked histogram or Pareto chart that shows the most common types of spills and leaks.
   b. Conduct a root cause analysis of the most common type of spill or leak.
3. PLAN: Develop a comprehensive corrective and preventative action plan to address the root causes, considering both short-term and long-term actions.

4. DO: Implement corrective and preventative actions to address the root causes.
   a. Issue one-point lessons to transfer best practices regarding the prevention of spills and leaks.
   b. Repeat the process with the next most common type of spill or leak.

4.4 AM Step 3: Tentative AM and SHE

AM Step 3, tentative AM, involves the development of provisional AM standards on equipment cleaning, lubrication, tightening, and checking designed to maintain basic equipment conditions.[10] Behavioral standards are developed in order to conduct cleaning, lubrication, tightening, and inspection in an efficient manner. The goal is to improve equipment reliability, maintainability, and safety by developing easy-to-understand standards. As the AM team develops standards, there is value in incorporating safety into the standards and systematically developing visual controls that will enhance equipment and plant safety.

4.4.1 Including SHE in Provisional Standards

Because SHE is not a separate activity but must be integrated into everyday work activities, safety, health, and environment issues should be included in tentative autonomous maintenance standards developed during AM Step 3. An effective way to communicate and emphasize the SHE information in cleaning, lubrication, tightening, and checking standards is to print it in a bold, distinctive font or highlight it in a brightly colored manner.

4.4.2 Use of SHE Visual Indicators and Controls

Although visual indicators provide some useful information, visual controls make workplace information and procedures self-explanatory and self-regulating by making the correct decision and correct action obvious to all workers. Visual controls provide clear information that directs the everyday actions of workers. Ideally, visual controls promote effective communication, effect the desired workplace action, and prevent worker errors. In addition, visual controls are useful for exposing production abnormalities that could result in losses or accidents. Thus, production visual controls increase process efficiency, effectiveness, and safety by reducing errors and ensuring that procedures get done one way—the right way. When implementing Lean and AM Step 3 a focus on developing and implementing SHE visual controls can further improve safety, health, and environmental performance. The goal should be to establish what Michel Greif terms a "visual factory"[11] where all key production and SHE information is simply and clearly communicated by visual means. Greif maintains that "visual communication is the predominant mode of communication within organizations that seek to reinforce employee autonomy."[12] SHE visual controls do not hide safety, health, and environmental information in memos and manuals, but instead put it where it is needed—in the *gemba* or workplace. Ideally, cross-functional teams of workers should develop agreed-upon ways to visually communicate important safety information. Useful SHE visual control methods include

1. Safety Walkways: As shown in Figure 4.4, brightly colored floor lines or markings can be used to indicate the location of plant pedestrian walkways. These walkways separate pedestrians from mobile equipment and hazardous processes, thereby enabling them to traverse safely from one area to another. The visual marking of walkways also serves to inform operators of mobile equipment where they can expect pedestrian traffic.
2. Machine Hazard Labeling: Standardized labels with distinct pictographs, as shown in Figure 4.5, can be used to mark the location of specific machine hazards such as pinch points, in-running nip points, sharp surfaces, shock hazards, and hot surfaces. Many injuries occur where

Figure 4.4   Safety walkways: a visual method of separating plant pedestrians from hazardous operations.

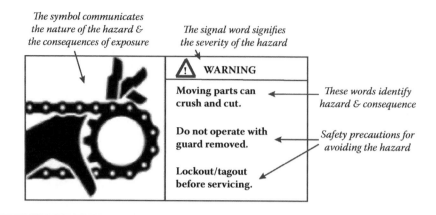

The symbol communicates the nature of the hazard & the consequences of exposure

The signal word signifies the severity of the hazard

⚠ WARNING

Moving parts can crush and cut. ← *These words identify hazard & consequence*

Do not operate with guard removed. ← *Safety precautions for avoiding the hazard*

Lockout/tagout before servicing.

Figure 4.5   Machine hazard communication via labels with visual pictograph.

employees' work activities bring them close to the working zone of equipment, thus it is important to place the warning labels in close proximity to the hazard. In this way, all site personnel are warned of potential machine hazards and can take measures to avoid them. Ideally, labels and signs should clearly communicate the following four critical pieces of safety information:

a. The nature of the hazard (the type of hazard; for example, crush, shock, cut, burn, etc.)

b. The consequence of exposure to the hazard

c. How to avoid the hazard

d. The severity of the hazard

The American National Standards Institute's (ANSI) Z535 series of standards provides the specifications and requirements for standardizing the format, color coding, symbols, and wording used on facility safety and environmental signs in the following documents:

ANSI Z535.1: Safety color code

ANSI Z535.2: Environmental and facility safety signs

ANSI Z535.3: Criteria for safety symbols

ANSI Z535.4: Product safety signs and labels

ANSI Z535.5: Safety tags and barricade tapes (for temporary hazards)

3. Machine Safeguard Color-Coding: Machine guards and safety devices can be painted or labeled in a bright color, such as yellow, so that operators are warned that a machine hazard is present in a particular area of the equipment. Standardizing the color-coding of machine safeguards, such as guards and interlocks, aids in their identification and facilitates the completion of equipment safety checks. As shown in

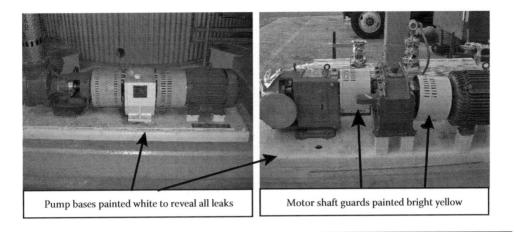

| Pump bases painted white to reveal all leaks | Motor shaft guards painted bright yellow |

Figure 4.6   Visual control applied to pump motor guards and pump bases.

Figure 4.6, painting the guard surrounding the rotating shaft of a pump motor yellow clearly communicates the hazard and makes it easy to identify when the guards are missing.

4. Pump Base and Containment Dike Painting: As shown in Figure 4.6, the bases of pumps and containment systems should be painted white so that a leak or spill can be quickly identified and addressed. To achieve high levels of safety and environmental performance, spills and leaks should not be considered an inevitable and acceptable part of conducting production operations. Concealing discharges with dark-colored surfaces is not the path to zero spills and leaks, but rather, small spills and leaks must be identified promptly and the root cause identified and corrected.

5. Machine E-Stops: Machine emergency stop buttons (Figure 4.7) should be of a uniform shape and color so that they can be easily and quickly identified in case there is a need to quickly stop the machine. In addition, each E-stop should be clearly labeled to indicate what equipment it deactivates. According to NFPA 79, the actuators of emergency stop devices should be colored red, and the background immediately around pushbuttons and disconnect switch actuators used as emergency stop devices should be colored yellow. The actuator of a pushbutton-operated device should be of the palm or mushroom-head type.

6. Color-Coded Indicator Lights and Standard Alarms: The color of each type of indicator light and the sound emitted by each equipment and emergency alarm should be standardized so that the warning can be readily identified and understood by all facility personnel. When an alarm

Figure 4.7 Visual color-coding and labeling of machine E-stops. (E-stop diagram courtesy of Banner Engineering, www.bannerengineering.com)

sounds or an indicator light is activated it should be readily apparent what the emergency is and what equipment is involved. Figure 4.8 provides an example of a light tree that provides visual warnings to operators of a process upset.

7. Machine Status Lights: The color, labeling, and shape of machine on–off indicators should be standardized throughout a facility. Typically, "on" indicator lights should be green and "off" indicators should be red. This type of ergonomic standardization prevents errors in which an operator incorrectly activates equipment that should be shut off. Figure 4.9 shows a novel machine status indicator that segments the indicator into four separate sections with different colors signifying four different machine states or conditions. Individual color segments can be lit separately or in combination to communicate a variety of machine conditions. These types of indicator lights can be used to improve Andon status reporting systems, operational error-proofing, and machine safety.

8. Energy Disconnects: If equipment energy disconnects are located a far distance from the point where employees perform their work tasks, they are likely to reach into the working zone of the machine without properly de-energizing the equipment. It is not realistic to expect workers to travel long distances to a remote motor control center, electrical panel, or valve in order to shut off and secure equipment via lockout and tagout. Close proximity energy disconnects are main energy

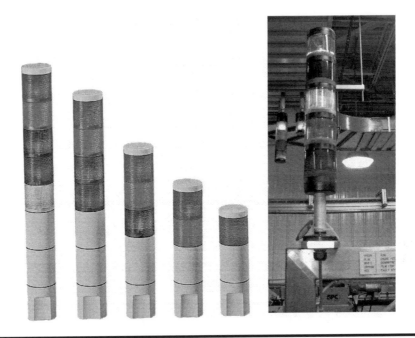

Figure 4.8   Alarm light tree with color-coded indicator lights. (Photo on left courtesy of Federal Signal Corporation© 2010. Used by permission.)

Figure 4.9   Machine status light with segmented indicator. (Photo of EZ-LIGHT™ Segmented Indicator courtesy of Banner Engineering, www.bannerengineering.com)

shutoffs that are positioned close to the point of operation, and thereby facilitate proper equipment de-energization and ensure employee safety. Using visual factory techniques, main energy disconnects should be brightly colored and labeled so operators can easily identify the location of the main energy shutoff and know exactly what equipment the disconnect de-energizes (Figure 4.10).

9. Pipe Color-Coding and Labeling: A color and labeling scheme should be developed to visually indicate the contents of all plant piping. For uninsulated pipes the entire pipe can be painted a unique color, and for insulated pipes uniquely colored labels and directional arrows can be used to designate specific pipe contents and direction of flow. ASME (ANSI) Standard A13.1, "Scheme for the Identification of Piping Systems,"[13] provides a general guide for pipe color-coding and marking. The lettering color-coding system used in this standard is summarized in Table 4.1. The maze of piping shown in Figure 4.11 provides a compelling example of why clear piping identification and unambiguous indications of flow direction are so important to ensure the efficient and safe operation of a complex process.

10. Tank and Vessel Labeling: Sites should develop a uniform approach for labeling the contents of tanks and vessels, and communicating the material hazards. As shown in Figure 4.12, large lettering that can be seen

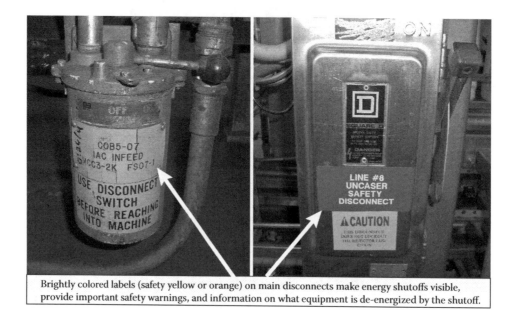

Brightly colored labels (safety yellow or orange) on main disconnects make energy shutoffs visible, provide important safety warnings, and information on what equipment is de-energized by the shutoff.

Figure 4.10 Visual techniques applied to main energy disconnects.

Table 4.1   Scheme for the Identification of Piping Systems

Material Properties/Classification	Color Scheme (Letter Color on Field Color)
Flammable Fluids	Black on Yellow
Combustible Fluids	White on Brown
Toxic and Corrosive Fluids	Black on Orange
Fire-Quenching Fluids	White on Red
Other Water (potable, cooling, etc.)	White on Green
Compressed Air	White on Blue
Defined by User	White on Black
Defined by User	Black on White
Defined by User	White on Purple
Defined by User	White on Gray

Source: ASME (ANSI) Standard A13.1, "Scheme for the Identification of Piping Systems."

Pipe labels provide information on contents, flow direction, and source

Figure 4.11   Visual pipe labeling.

from a distance should be used to indicate the tank contents, and an NFPA or similar warning label can be used to communicate the material hazards. The NFPA 704 labeling system uses a diamond-shaped symbol divided into four color-coded quadrants to communicate the hazards of a material and the degree of severity. The health hazard rating is

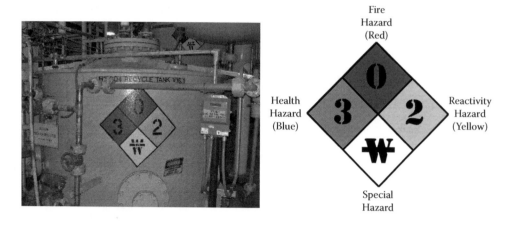

Figure 4.12 Visual tank and vessel labeling.

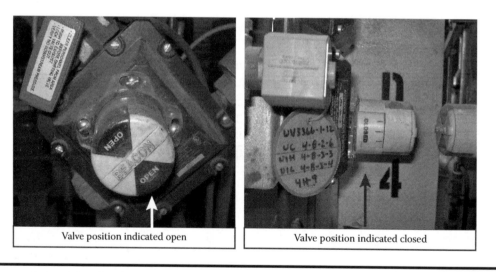

Figure 4.13 Visual valve position indicators.

given in the blue quadrant, flammability in the red section, reactivity in the yellow, and special hazards listed in the white portion of the label. Hazard severity is indicated by a numerical rating that ranges from zero (0) indicating a minimal hazard, to four (4) indicating a severe hazard.[14]

11. Valve Position Indicator: Visual devices, as shown in Figure 4.13, can be installed on valves to indicate whether the valve's current position is open or closed. In addition, colored tags can be placed on valves to indicate if the valve's normal position is open (green) or closed (red). Additional information can be placed on the tag such as the valve number and service.

12. Visual Gauge Indicators: The face of a gauge can be color-coded to visually indicate the proper operating zone for the process, and where it is unsafe or undesirable to operate the process. When gauges are clearly marked it enables both operators and other plant personnel to quickly identify if equipment or a process is operating outside the desired limits. Marked gauges enable even persons unfamiliar with the equipment to identify operating abnormalities, thereby increasing the number of eyes that are monitoring the process. Figure 4.14 demonstrates several methods for marking gauges.

13. Height Limit Indicators: Lines and tell-tale indicators are posted to warn operators before they exceed a safe maximum height level. For example, as demonstrated in Figure 4.15, a warning line or chains can be hung to signal truck operators that their vehicle height exceeds that of a low overhanging structure they are approaching. Thus the vehicle can be stopped and prevented from affecting the structure.

14. Tool Shadow Boards: As stated before, a key principle of Lean is "a place for everything and everything in its place." When plant equipment, materials, and supplies are stored properly, facility housekeeping, safety, and efficiency are improved. In addition, it is important that work tools be properly stored and easily retrievable so that time is not wasted hunting for the right tool, and to prevent employees from using the wrong tool

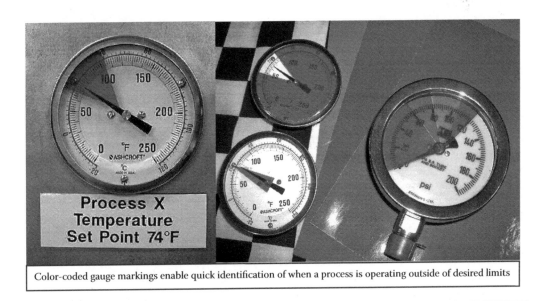

Color-coded gauge markings enable quick identification of when a process is operating outside of desired limits

Figure 4.14 Visual gauge markings. (Photo courtesy of Strategic Work Systems, Inc., www.leanmachines.com)

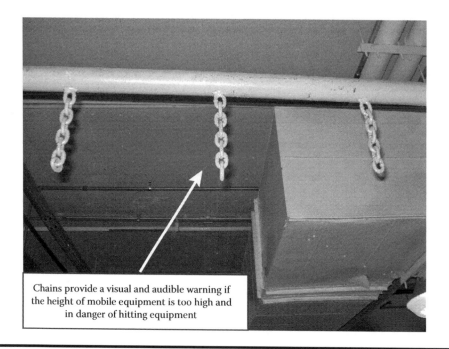

Chains provide a visual and audible warning if
the height of mobile equipment is too high and
in danger of hitting equipment

Figure 4.15 Visual maximum height indicators.

Figure 4.16 Tool shadow boards.

for the job. As shown in Figure 4.16, shadow boards that have outlines
of each tool provide a visual indication of whether a tool has been mis-
placed, and thereby increase the probability that the proper tool will be
available when needed.

15. Emergency Equipment and PPE Storage: Emergency Personal Protective
Equipment (PPE) such as escape masks and self-contained breathing apparatus (SCBAs) should be immediately available at the point
of operation. Standardized, brightly colored storage cases make the

location of this important safety equipment readily visible in the event of an emergency and protect it from damage by the environment. Likewise, brightly colored cases or boards can be used to store lockout/tagout devices in a single, readily accessible place near the point of use (Figure 4.17).

16. Lockout/Tagout Diagrams and One-Point Lessons: Prior to performing a task, employees are less likely to read long and wordy procedures than to review a visual safety diagram or one-point lesson. In light of this, lockout/tagout diagrams should be posted near equipment to visually communicate the proper procedure for de-energizing and securing equipment. Safety OPLs should also be hung near the affected equipment to visually communicate the safe way to perform a task.

17. Personal Protective Equipment (PPE) Pictographs: A picture is worth a thousand words. Rather than using lengthy wording on safety signs, pictographs can be used to more effectively communicate personal protective equipment requirements for a particular work area or task. Figure 4.18 shows how pictographs can be used on signs to make it crystal clear what the PPE requirements are for a task in which hazardous materials are handled.

18. Safety Shower and Eye Wash Marking: The location of all site safety showers and eye washes can be communicated by means of signs and green boxes painted on the floor beneath the shower. The green box serves to visually indicate the location of the emergency equipment and warns employees not to impede access to the shower by storing anything below it (Figure 4.19). Because time is critical when one is exposed to a

Figure 4.17   Visual emergency equipment and PPE storage.

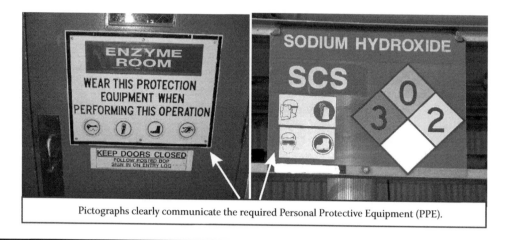

Pictographs clearly communicate the required Personal Protective Equipment (PPE).

Figure 4.18 Personal protective equipment (PPE) pictographs.

Figure 4.19 Visual marking of safety showers and eye wash stations.

hazardous material, visual marking of safety shower and eye wash stations that makes the locations stand out can prevent serious injuries.

19. Fire Protection Equipment Marking: As shown in Figure 4.20, red boxes can be painted on the floor below fire protection equipment, such as fire extinguishers and fire hoses, to visually communicate its location and to warn persons not to hinder access to the equipment by storing material under it or in front of it.

Figure 4.20 Visual marking of fire protection equipment.

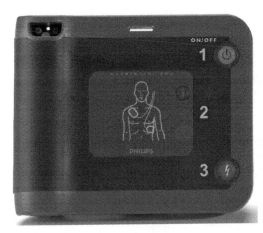
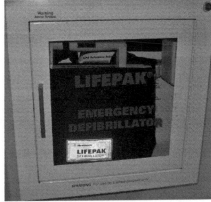

Figure 4.21   Visual AED storage and directions for use. HeartStart FRx AED. (Photo courtesy of Philips Healthcare.)

20. AED/Medical Supply Stations: Automatic external defibrillators (AEDs) and medical supplies should be stored in brightly colored and labeled storage cases so that they can be quickly located in case of an emergency (Figure 4.21). Time is crucial in a medical emergency, thus quick identification of the location of medical supplies is important. In addition, illustrations and pictures can be used to clearly communicate the proper operating procedure for the AED.

21. MSDS/Right-to-Know Stations: Material safety data sheet (MSDS) or right-to-know stations should be marked in a distinctive fashion with

Figure 4.22 Visual marking of MSDS right-to-know station.

signs and wall markings. Figure 4.22 shows an MSDS station with distinctive diagonal green stripes marking where MSDSs are available in both hardcopy and electronic form.

22. Standard Spill Warning Markers: Sites should develop a standard that requires all spills to be marked immediately and in a uniform fashion. Brightly colored cones or signs can be used to mark the location of a spill and to warn workers against entering the spill area. Figure 4.23 shows spill cones that can be customized with warning legends and flashing lights.

23. Spill Supplies: Spill supplies should be stored in a secure, standardized, and labeled container so that they are protected from the elements and can be quickly located in the event of an emergency. The contents of

Figure 4.23   Safety cones with visual legends and lights for marking spills and leaks. (Photo courtesy of Jackson Safety, Division of Kimberly-Clark Worldwide, Inc.)

the spill kit or container should be listed on the container so that the supplies can be periodically inventoried and verified. Figure 4.24 shows a spill cart with all supplies labeled and arranged for quick access.

24. Drain and Exhaust Stack Color-Coding: Facility drain covers should be painted unique colors to designate the type of effluent discharge. For example, green drain covers can signify storm drain discharges, and yellow manholes can be used to designate sanitary sewer discharges (Figure 4.25). In addition, a map of a facility's effluent discharge system can be developed where each drain line is numbered and labeled to indicate which process discharges to a particular line. In this way, the source of spills or elevated effluent discharges can be quickly identified and corrected. In a similar fashion, a facility's emission points or exhaust stacks can be numbered, color-coded, and labeled to enable prompt identification of the source of elevated air emissions.

25. Visual Marking of Electrical Disconnects and Panels: As shown in Figure 4.26, electrical disconnects or shut-off devices should have a uniform color and should be labeled with information on the electrical source, electrical shock, and arc flash hazards, as well as storage requirements. Brightly colored lines can be placed in front of electrical panels to prevent storage in this area from blocking access to the panels.

Figure 4.24 Supplies labeled and arranged in a spill cart for quick access.

SANITARY SEWER
DRAIN

STORM SEWER DRAIN
TO CHESAPEAKE BAY
NO DISCHARGE

Figure 4.25 Visual effluent drain marking.

26. Flammable Liquid Storage: Ideally, all flammable liquid storage cabinets used at a site are similar in color and appearance. Boxes can be painted on the floor, marking the designated location of the cabinet and thereby ensuring adequate air flow through the cabinet vents and easy access to the cabinet (Figure 4.27). The options for applying visual indicators and visual controls to the safety, health, and environmental area are numerous.

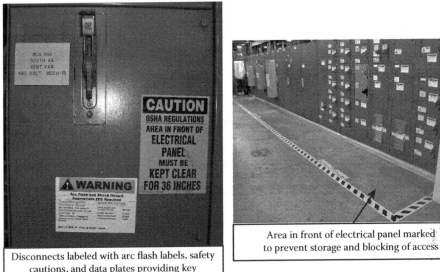

Disconnects labeled with arc flash labels, safety cautions, and data plates providing key information on electrical sources & equipment served

Area in front of electrical panel marked to prevent storage and blocking of access

Figure 4.26   Visual marking of electrical panels.

Figure 4.27   Flammable liquid storage.

The site SHE pillar team should challenge area teams to develop visual means to improve the communication of SHE hazards, error-proof SHE procedures, and reduce SHE accidents, incidents, and losses. In the end, well-designed visual controls make operations easier, more efficient, and safer.

4.4.3 Pointing and Naming Drills

In AM Step 3 operators develop provisional standards and checklists on equipment cleaning, lubrication, tightening, and inspection. In a similar way inspection standards and checklists can be developed for checking key safety, health, and environmental parameters of the process. In pointing and naming drills operators complete a SHE checklist one step at a time by pointing to the relevant SHE item, naming it, and checking it. The practice of pointing and naming ensures that every step of the checklist is covered and develops the discipline to visually check each item thoroughly. Just as an airline flight crew uses a checklist to ensure that the plane is ready to safely take off, a production line crew can conduct a naming and pointing drill to complete a manufacturing-line checklist to ensure the equipment's safety and readiness for operation.

The first three steps of autonomous maintenance focus on eliminating forced equipment deterioration and establishing and maintaining basic equipment conditions. The purpose and philosophy of the three initial AM Steps (initial clean, countermeasures, tentative AM) can be summarized as, "Clean to inspect, inspect to detect, and detect to correct."[15] As a result of the autonomous maintenance activities of Steps 1 through 3, equipment is made more reliable, operators are made more knowledgeable, and the process is made safer. By ensuring that SHE is integrated into each AM step, SHE can be fully and effectively integrated into the organizational culture.

4.5 AM Step 4: General Inspections and SHE

In AM Steps 4 and 5 the focus shifts from preventing equipment deterioration to providing operators with the knowledge and skills that will enable them to work more effectively. During AM Step 4, general inspection, operators receive comprehensive training that gives them an in-depth understanding of the equipment's principles of operation, its function, and its design and structure. With this increased knowledge workers conduct routine general equipment inspections and adjustments including visual checks, bolt tightening, calibrations, and replacement of worn parts.[16] According to Tokutaro Suzuki, "Developing equipment-competent operators revolutionizes not just equipment management but every other aspect of workplace management as well."[17] Operators are able to prevent equipment failures and improve safety through improved inspection and monitoring, and by performing basic equipment maintenance. By incorporating detailed SHE training into the AM Step 4 curriculum, workers develop an understanding of equipment hazards and how to avoid incidents and injuries. Figure 4.28 lists some of the SHE activities undertaken during AM Steps 4–7.

4.5.1 Incorporating Real-Life Safety Examples into AM Training

As line operators are trained on their equipment and how to perform AM activities it is important to incorporate safety, health, and environment into the training. Training of operators on specific SHE cases using real plant examples will make the training more meaningful, memorable, and effective. Some plants establish an area in the plant known as the "Safety Gym" where hands-on safety displays, equipment mock-ups, and demonstrations of actual accidents and safety-related equipment failures are used in the training. For example, mock-ups of equipment energy sources and methods of machine de-energization and lockout/tagout will provide operators with key knowledge and skills needed to work safely with equipment. These real-life examples provide insight into the root cause of safety incidents and the proactive actions needed to prevent them from reccurring. The emphasis is on making safety, health, and environmental training memorable and relevant to the work situation.

4.6 AM Step 5: Autonomous Inspection and SHE

In AM Step 5, autonomous inspection or checking, improvements are made to provisional AM standards to make cleaning, lubrication, and inspections more efficient. To enable operating personnel to take full responsibility for the operation and inspection of their equipment, checklists, comprehensive visual management, and error-proofing techniques are emphasized to eliminate mistakes that can result in downtime and accidents.[18]

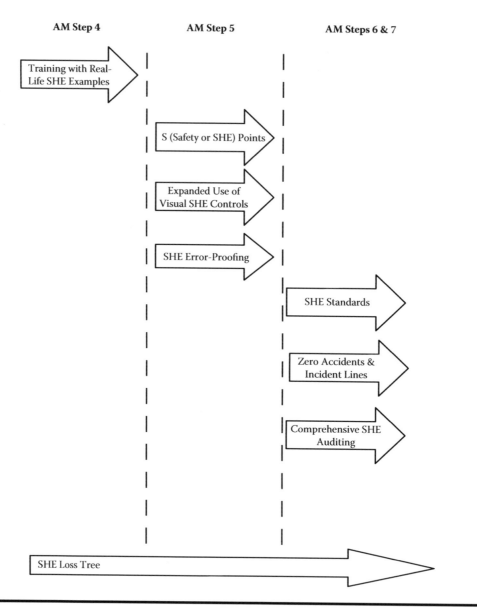

Figure 4.28 SHE activities implemented during AM Steps 4–7.

4.6.1 S Points (Safety or SHE Points)

The visual marking of key equipment safety or SHE points that need to be checked on a regular basis can be incorporated into the AM inspection program. These safety points can include items such as critical safety set-points, equipment interlocks, machine guarding, alarms, and equipment de-energization devices. The safety or S points can be numbered and marked with brightly colored labels to ensure their easy identification during inspections. In some facilities a standardized inspection path is established by placing symbols of shoes on the floor marking the exact location of the inspection point. In this way, inspection points can be easily identified and all operators are positioned at the same spot when performing the inspection.

4.6.2 SHE Mistake-Proofing or Poka Yoke

Safety mistake-proofing, or *poka yoke* in Japanese, involves the use of mechanisms to prevent safety errors from occurring or to detect the potential mistake and enable corrective action before an accident occurs. Safety mistake-proofing involves designing, engineering, or modifying a process to eliminate the possibility of human error. Multidisciplinary teams of operators, maintenance personnel, engineers, and safety professionals can be established to brainstorm *poka yoke* approaches to make the process safer by eliminating errors that can cause accidents and incidents. Ideally, mistake-proofing methods should be simple to apply and relatively inexpensive. Examples of mistake-proofing approaches that enhance safety include

- Designing chemical transfer hoses with unique fittings so that they cannot be inadvertently connected to the wrong tank or process
- Different colors and connections on oxygen and nitrogen lines to prevent inadvertent cross-connections and the resulting asphyxiation if nitrogen were used instead of oxygen, or explosions if oxygen were pumped into a vessel that required inerting
- Trapped key interlocks for ensuring safe de-energization and lockout/tagout of equipment
- Surgical instrument trays with specific indentations for each tool ensuring that no tool is left inside the patient[19]

4.7 AM Step 6: Standardization and SHE

The AM standardization step aims to reduce variation and impose process discipline through the development of standards that clearly specify work area procedures on housekeeping and work flow.[20]

4.7.1 SHE Standards

As part of AM Step 6, effort is devoted to reducing variation in the implementation of safety and environmental procedures. A multidisciplinary small group team is set up to analyze safety and environmental procedures with the aim of improving, optimizing, and standardizing the SHE tasks. Work-flow diagrams, illustrations, and videotapes of activities can be used to aid the evaluation of SHE procedures. When documenting the agreed-upon optimized procedure, it is recommended that visual methods be utilized as much as possible to convey the correct way to perform the task.

4.7.2 Zero Accident and Zero Incident Lines

By Step 6 of autonomous maintenance, the Lean philosophy and the associated equipment, people, and work system improvements have become embedded in many areas of the organization. Some production lines and work areas have demonstrated that achieving zero accidents, zero incidents, and zero losses is no longer a dream but a reality. The leading practices of zero accident and zero incident lines should be identified and deployed to all work areas. Leading SHE practices should be standardized and replicated throughout the site by incorporating them into site SHE standards.

4.8 AM Step 7: All-Out AM Management and SHE

The purpose of AM Step 7 is to consolidate the equipment improvements, reinforce employee development, and institutionalize the work system changes that have been achieved through the preceding autonomous maintenance steps. The goal is to establish a self-directed work system and culture where employees are empowered and actively manage all aspects of the production process. Self-directed teams assume responsibility for equipment and work area improvement, and for the safety of their work area.

4.8.1 Comprehensive SHE Inspection and Auditing

Under TPM and Lean, small-team autonomous management of the work area includes ensuring the overall effectiveness of both the production process and the safety, health, and environmental management system. By AM Step 7 operator teams assume control for identifying and solving production and safety problems. One way in which the work area team can take ownership for SHE and identify and solve safety issues is to become actively involved in the SHE inspection and auditing processes.

Operator safety inspections should involve regular, comprehensive safety checks of the conditions and work practices in their work area. Using detailed checklists operators check key safety items, safety points (S points), and work practices in their work area. As previously discussed under the 6S or 5S for SHE process, these inspections should involve the systematic identification and control of all potential safety hazards. By AM Step 7 the scope of workplace safety inspections has expanded upon the 5S for SHE process's focus on basic area conditions to include all equipment and area safety issues identified in the previous AM steps, and on critical safety behaviors. These comprehensive area safety inspections should employ the same visual control and problem-solving methods utilized in the 5S for the SHE process. Safety and environmental hazards that are identified are photographed and SHE F-tags are completed for each item. Activity boards, measles maps, and Pareto charts are developed to visually communicate the location and types of safety and environmental hazards and the methods for controlling them. A standardized problem-solving methodology such as CAPDo or PDCA is employed to efficiently identify and control the hazards.

Traditionally, workplace SHE auditing is a management role and responsibility. In AM Step 7 operator teams are actively involved in coordinating and conducting audits of their area's safety, health, and environmental management systems. In many sites the internationally recognized safety and environmental standards such as OHSAS 18000 and ISO 14000 serve as the basis for developing and assessing safety, health, and environmental management systems. With the guidance and assistance of safety, health, and environmental professionals area operators are trained on the organization's SHE management systems, and how to audit the key elements of the SHE management system.

Under the ISO 14001 model[21,22] and OHSAS 18001[23] an environmental management system (EMS) and a health and safety management system (HSMS) each have 17 similar elements. Many organizations integrate their EMS and HSMS into a unified safety, health, and environmental management system with the following elements:

1. SHE policy
2. Health and safety hazards and environmental aspects identification

3. Legal and other requirements
4. SHE targets and objectives
5. SHE management program
6. Structure and responsibility
7. Training, awareness, and competence
8. Communication
9. Documentation
10. Document control
11. Operational control
12. Emergency preparedness and response
13. Monitoring and measurement
14. Nonconformance and corrective and preventive action
15. Records
16. SHE management system audit
17. SHE management review

Using standardized checklists, operator teams conduct an annual self-assessment of each SHE element. Using the checklists, key SHE gaps are identified and action plans are developed to close these gaps. To make the self-assessment process manageable by the operator teams it is often easier to assess one or two SHE elements per month rather than auditing all of them at one time. Each SHE management system question can be assigned a score and the results can be communicated visually via a radar or spiderweb chart.

Autonomous maintenance is the pace-setting pillar of TPM and Lean. The success in implementing autonomous maintenance influences the progress of all other pillars and the ultimate success of the overall TPM and Lean initiative. When implemented in a planned, systematic, and thorough manner, autonomous maintenance results in continual improvement of equipment conditions, employee capabilities, process performance, and workplace safety, health, and environment. It is important to understand that because each step of autonomous maintenance builds upon the previous step, it is essential to implement each step thoroughly without rushing to the next step. To derive the most safety, health, and environmental benefit from the implementation of autonomous maintenance it is crucial to apply the specific tools and methodologies of each AM step to the organization's SHE process. Safety, health, and environment are not managed as a separate process but are fully integrated into the organization's

Lean process and way of working, thus the vision of zero losses, zero accidents, and zero incidents can be realized.

Endnotes

1. Suzuki, Tokutaro (1994). *TPM in Process Industries.* Productivity Press, New York, pp. 102–103.
2. Robinson, Charles and Andrew Ginder (1995). *Implementing TPM—The North American Experience.* Productivity Press. Portland, OR, pp. 93–117.
3. Suzuki, op. cit., p. 326.
4. Suzuki, op. cit., pp. 328–329.
5. Suzuki, op. cit., pp. 101–102.
6. Nagamachi, Mitsuo (2008). SHIROMISO-AKAMISO accident prevention management based on brain theory. Hiroshima International University. *2008 International Conference on Productivity & Quality Research.*
7. Suzuki, op. cit., pp. 106–108.
8. Leflar, James A. (2001). *Practical TPM: Successful Equipment Management at Agilent Technologies.* Productivity Press, Portland, OR, pp. 78–82.
9. Soin, Sarv Singh (Sept., 1998). *Total Quality Essential: Using Quality Tools and Systems to Improve and Manage Your Business.* McGraw-Hill Professional, New York.
10. Suzuki, op. cit., pp. 114–118.
11. Greif, Michel (1991). *The Visual Factory.* Productivity Press, Portland, OR.
12. Greif, op. cit., p. 14.
13. ASME (ANSI) A13.1-2007 (2007). *Scheme for the Identification of Piping Systems.* American Society of Mechanical Engineers, New York.
14. NFPA 704-2007 (2007). *Standard System for the Identification of the Hazards of Materials for Emergency Response.* National Fire Protection Association. Quincy, MA.
15. Smith, Rick L. (Sept., 2004). "All breakdowns can be prevented." *Fabricating and Metalworking Magazine.* Retrieved on September 21, 2009 from: http://www.technicalchange.com/pdfs/why-are-these-men-smiling.pdf
16. Robinson and Ginder, op. cit., pp. 107–109.
17. Suzuki, op. cit., p. 119.
18. Robinson and Ginder, op. cit., pp. 109–112.
19. Chase, R. B. and D. M. Stewart (1994). Make your service fail-safe. *Sloan Management Review.* (Spring): 35–44.
20. Robinson and Ginder, op. cit., pp. 112–116.
21. Martin, Raymond (1998). *ISO 14001 Guidance Manual.* National Center for Environmental Decision Making Research. Technical Report NCEDR/98-06. Retrieved on January 20, 2010 from: http://www.usistf.org/download/ISMS_Downloads/ISO14001.pdf

22. ISO (2004). *ISO 14001:2004 Environmental Management Systems—Requirements with Guidance for Use.* International Organization for Standardization. Geneva.

23. OHSAS (2007). OHSAS 18001-2007: Occupational Health and Safety Management, Requirements. British Standards Institution (BSI). London, UK. July 31. Available from: http://shop.bsigroup.com/ProductDetail/?pid=000000000030148086

Chapter 5

Integrating SHE, Planned Maintenance, and Early Management

The worst thing is to rush into action before the consequences have been properly debated. … We are capable at the same time of taking risks and estimating them beforehand.

Pericles' funeral oration
(in Thucydides' History of the Peloponnesian Wars circa 430 BC) Quoted in Kletz, 1999[1]

5.1 Rationale behind the Integration of Planned Maintenance and SHE

The basic objective of the planned maintenance or PM pillar (*Keikaku Hozen* in Japanese) is to eliminate equipment failures and breakdowns and thereby achieve uninterrupted and optimal production.[2] The PM pillar aims to establish and implement an effective maintenance strategy for facility equipment and systems that increases their reliability and safety, provides high service levels throughout the life of the assets, and maximizes the benefit of limited maintenance resources. Effective integration of the PM and SHE pillars is essential to achieving these goals of the PM pillar, and to the ultimate success of an organization's TPM and Lean initiative. Application of the SHE pillars' systematic approach to hazard recognition, assessment, and control to maintenance tasks is vital to safeguarding site personnel and ensuring the effective utilization of the organization's limited skilled maintenance resources.

By integrating the TPM SHE and planned maintenance or effective maintenance pillars, an organization will reduce the risk of injury and environmental incidents by

1. Preventing equipment failures that could result in accidents or incidents
2. Reducing the exposure of maintenance and operating personnel to unreliable and unsafe equipment
3. Improving implementation and control of the equipment-servicing activities of site personnel
4. Reducing operator errors through the transfer of equipment knowledge from skilled maintenance staff to operating personnel
5. Incorporation of equipment learning and best practices into autonomous maintenance procedures and improved equipment design
6. Ensuring the competence and qualification of all personnel who perform specialized and potentially hazardous equipment-related tasks such as welding, lifting heavy loads, electrical tasks, confined space entry, work at heights, refrigeration repair, pressure vessel operations, and so on

Certain inspection and maintenance activities lie beyond the scope of the autonomous maintenance performed by equipment operators, including

- Equipment assessment and repair tasks requiring specialized skills
- Overhaul repair in which deterioration is not visible from the outside
- Repairs to equipment that is hard to disassemble and reassemble
- Tasks requiring special measurements
- Tasks posing substantial safety risks, such as working in confined spaces or working at heights[3]

The PM pillar is concerned with ensuring that these tasks performed by maintenance specialists are conducted effectively, efficiently, and safely.

5.2 Safety Permit Systems

The efficient and safe execution of routine production and maintenance tasks is generally accomplished by means of detailed written job procedures or job safety analyses (JSAs) that outline the hazards and associated safety precautions for each step in a particular task. Many equipment maintenance tasks, however, are nonroutine in nature, and therefore do not have written procedures. Owing to their nonroutine nature, these jobs often present unique hazards, involve elevated risk, and require specialized knowledge in order to execute properly and safely. Therefore, to ensure that these high-hazard and nonroutine tasks are completed without injury or incident it is necessary to perform some type of risk assessment each time they are undertaken.

Typically, the risk assessments and hazard control methods for nonroutine maintenance tasks are documented by means of a formal safety work permit system. A safety work permit system (or permit to work system) is a documented management procedure in which authorized and trained personnel conduct a simplified risk assessment using a standardized permit form. Figure 5.1 provides an example of a general safety work permit form that includes checklists that serve as a written record of the risk assessment by specifying the hazards of the task and the necessary precautions to be taken to control them. The permit authorizes specific work at a specific location and at a specific time as long as the agreed safety precautions are taken. In leading safety organizations general safety work permits are commonly used to control the hazards presented by all nonroutine tasks. In addition, specialized safety permits are required for high hazard tasks such as: confined space entry, hot-work, line-breaking, demolitions, excavations, heavy lifting with cranes, work on high-voltage equipment, and working at heights.

The PM pillar in conjunction with safety experts and the SHE pillar should design and implement a site safety work permit system. An inventory of facility tasks and jobs requiring both general and specific safety permits should be developed to ensure full understanding of what activities will require a permit. Figure 5.2 is an example of the second page of an integrated safety work permit form that includes permits for several types of high-hazard, nonroutine maintenance activities on a single form. The checklists for each high-hazard task serve as a mental reminder to ensure that all necessary safety equipment is worn and the required safety precautions are taken.

5.3 Planned Maintenance Support for AM and SHE

Because much of an organization's knowledge and technical expertise regarding its equipment resides with its maintenance personnel, it is essential that craft personnel and the PM pillar team work closely with the autonomous maintenance and safety, health, and environmental pillars. The

Safety Work Permit & Risk Assessment

PERMIT MUST BE POSTED & IS VALID FOR 8 HRS OR UNTIL WORK CREW CHANGES

All contractor work requires completion of a permit & signatures.

| Contractor: | | Dept: | | Contact Information: | |
| Location: | | DATE: | | SHIFT / TIME: | |

Description of Work:

A. MANAGEMENT OF CHANGE CHECKLIST
Does the job, change result in any of the following risks?

RISK	yes	no	RISK	yes	no
1. Results in exposure to a hazardous material (ie: ammonia, HCL, sulfuric acid, caustic, nitrogen)?			2. Inactivates or reduces the efficiency of an emergency system (ie: alarms, Scrubbers , fire protection...)?		
3. Exceeds safe operating limits in operating conditions?			4. Inactivates or alters a process safety interlock?		
5. Inactivates or alters an alarm or its set point?			6. Inactivates or alters a pressure relief device?		
7. Increases the possibility of a chemical release?			8. Causes material of construction or piping changes?		
9. Violates an emergency code or safety standard?			10 Involves the use of a new hazardous chemical?		

● If a yes answer is given to any question then a detailed Process Hazard Analysis/Safety Review must be conducted, and a formal written Process Change Authorization must be obtained.

For wiring or programming changes, you must provide additional information

B. POTENTIAL HAZARDS	Hot Work		Flammable Atmosphere	Hot Surface	Sharp Surfaces	Slip/Trip
	Confined Space		Oxygen Deficiency	Engulfment	Toxic	Asbestos
	Hazardous Energy		High Noise	Radiation	Chemicals:	Lead Paint
	Work at Heights		Fall Hazard	Lifting	1. 2.	

C. LIST THE PPE & EQUIPMENT REQUIRED
In addition to the standard PPE (Safety Shoes & Safety Glasses)

Splash Proof Goggles	Hearing Protection		Safety Constructed Scaffolds	Ground Fault Interrupter	Barricades
Face Shield	Respirator		Safety Locks and Danger Tags	Spark Proof Tools	Warning Signs
Rubber Coat or Suit	Hard Hat		Equipment Grounding Required	Personnel Carrier	Air Horn
Rubber Boots	Type:	Gloves	Low Voltage Extension Light	Confined Space Cart	Safety Harness

Addition Precautions:

D. SPECIAL SAFETY PERMITS REQUIRED

1. Work At Heights: Does the job require work at height (WAH) or an unprotected surface or require the use of an unapproved anchorage point?	**Applicable** *WAH SWP Required*	**Not Applicable**
2. Hot Work: Does the job require cutting/welding/brazing or any other type of heat/spark generating activity?	**Applicable** *Hot Work Permit Required*	**Not Applicable**
3. Confined Space: Does the job require entry into a confined space?	**Applicable** *Confined Space Permit Required*	**Not Applicable**
4. Line Breaking: Does the job require cutting into or in any way breaking a line or pipe?	**Applicable** *Line Break Permit Req.*	**Not Applicable**
5. Hazardous Energy: Does the job require working with hazardous energy for which there is no written procedure?	**Applicable** *Energy Control Procedure Required*	**Not Applicable**

Figure 5.1   Example of a general safety work permit.

PM pillar team should provide training, guidance, and coaching to operating personnel to enable them to gain the necessary knowledge and skills to manage their equipment efficiently and safely on a daily basis. By working together as a team the AM, PM, and SHE pillars endeavor to increase the reliability, effectiveness, and safety of the site's equipment.

E. ENERGY CONTROL PROCEDURE

		Applicable		Not Applicable	

HAZARDOUS ENERGY TYPE:	Electrical	Mechanical	Chemical	Pneumatic	Hydraulic	Temperature	Gravity
SOURCE:							

ISOLATION (SHUTOFF) DEVICE:		Safety Disconnect	Air Dump	Valve	Other	
LOCATION:						

LOCKOUT DEVICE:	Lock and Tag	Valve Chain or Cover	Other	

RESIDUAL ENERGY:	Yes	No

RESIDUAL ENERGY SOURCE:
Means to dissipate or restrain:
AFFECTED PERSONS TO NOTIFY (job titles):

Are you comfortable with all aspects of this procedure?	Yes	No

If no, STOP and talk with your supervisor

(***) A separate Energized Electrical Work Permit is required when <u>energized</u> electrical work is to be performed.

F. REVIEW & APPROVAL: All lines must be signed.

TECHNICIAN:
Signature:	Date:	Time:	

SUPERVISOR:
Signature:	Date:	Time:	

G. SPECIAL SAFETY WORK PERMITS

1. WORK AT HEIGHTS PERMIT: CHECK ALL THAT APPLY

	Applicable	Not Applicable
Full Body Harness and arrestor	Anchorage Point: (Must be verified by competent person)	
Guardrail	Ladder Tie-off	Spotter (if required)

2. CUTTING/WELDING/HOT WORK PERMIT

	Applicable	Not Applicable
Fire resistant barriers, screens used to confine heat, sparks and slag, if applicable	Combustible/Flammable material moved away from site	
Flammable Vapor Test satisfactory (LEL < 10%)	Fire Watch provided, if applicable	Fire Extinguisher available

Gas Test Report:	LEL %	DATE	TIME(S)	Gas Tester Signature:

3. LINE BREAKING PERMIT

	Applicable	Not Applicable

Possible material(s) in the line:

Precautions Checklist:	Type of Line to be Broken:		
		Description	Approvals
Has the area been roped off and warning signs posted?			
Have the other necessary areas of this form been completed?	Chemical		
Has all necessary PPE been made available?			Dept. Supv.
Have the Pumps & other energy sources been tagged & locked out?	Product		
Have the Valves been properly positioned, tagged, and locked out?			Dept. Supv.
Have the Lines been properly drained?	Sanitary Sewer		
Has the residual pressure been bled?			Environmental
Have the necessary blanks been installed?	Storm Sewer		
Has the Steam or Heat Tracing been turned off?			Environmental
Have both sides of the line been adequately supported?	Potable Water		
Have lead abatement/precautions been undertaken?			Environmental
Have the necessary lines been disconnected?	Other		
Has approval been received for connecting the new lines?			Environmental

4. CONFINED SPACE ENTRY PERMIT

	Applicable	Not Applicable
Adequate ventilation to maintain a safe atmosphere while permit is in effect.	Toxic gas test satisfactory, if applicable (H_2S, SO_2, CO, VOCs)	
Adequate illumination provided by lanterns or six-volt extension lamps	Flammable Vapor Test satisfactory (Reading < 1% investigate all readings 5% of LEL)	
All pipelines blanked at connections. (N_2, steam, chemical)	Oxygen Deficiency Test satisfactory (> 19.5%)	
Hazardous materials will not be introduced by work	Oxygen Enrichment Test satisfactory (<23.5%)	
Body harness and lifeline worn in confined space	Electrical controls are locked out and tagged	Electrical equipment grounded
Emergency response personnel notified	SCBA available	Trained attendant present NAME

Gas Test Report:

OXYGEN %	LEL %	TOXIC GAS CO	H₂S	VOC	TIME(S)	Gas Tester Signature:	
						Meter Model #:	
ENTRANTS: 1. 2.		TIME IN		TIME OUT	QUALIFIED?	ENTRY SUPV. SIGNATURE:	

Figure 5.2 Example of safety permits for high-hazard activities.

PM support for AM and SHE can be provided in the following specific areas:

1. Technical support in developing equipment lockout/tagout (LOTO) and de-energization diagrams and procedures
2. Conducting training sessions on the principles of machine operation
3. Assistance in the identification and delivery of machine-related safety training including machine safeguarding, machine noise and vibration control, machine risk assessment, and permit-to-work systems
4. Hands-on support and participation in equipment autonomous maintenance activities
5. Promptly addressing equipment abnormalities identified via maintenance F-tags
6. Providing technical guidance during the development of equipment autonomous maintenance and safety procedures
7. Participating in multifunctional equipment *kaizen* teams
8. Guidance on the care and maintenance of critical parts whose failure can result in major loss, including injury or environmental incidents
9. Guidance on the systematic inspection, calibration, and maintenance of safety and environmental systems
10. Assistance in the investigation and root cause analysis leading to corrective and preventive action related to incidents of equipment failure that have, or could have, resulted in injury or environmental incidents

5.4 PM Safety Checklists

Although operating personnel via autonomous maintenance are involved in the routine management and preventive maintenance of their equipment, craft personnel via planned maintenance are involved in advanced maintenance activities aimed at restoring, improving, and extending equipment's functioning lifetime. To achieve these equipment reliability aims, expert maintenance personnel are involved in periodic, predictive, and corrective maintenance. When conducting equipment checks and diagnostic tests, checklists are used to ensure that the tasks are performed uniformly and safely. When properly designed and utilized, checklists can be powerful tools for ensuring the safe and efficient operation of equipment. In fact, research has indicated that checklists are effective in reducing errors associated with complicated tasks in many fields including medicine, aviation, and industry.[4]

Another activity performed by craft personnel that has important safety implications is maintenance prevention—the process of reducing and ultimately eliminating maintenance through equipment and design improvements. By eliminating the need to conduct certain maintenance tasks plant personnel are removed from exposure to potentially hazardous tasks.

5.5 Early Management and SHE

Early management (EM) includes both early product management (EPM) and early equipment management (EEM) and involves a system of proactive activities conducted during the design, development, and launch of new products, processes, and equipment. These early management activities are aimed at designing error-proof and user-friendly equipment and products that meet and even exceed customer desires. By employing the methodologies and tools of early management a business can reduce development time and achieve a vertical start-up while reducing losses and producing equipment and products with optimal design and functioning.

Safety, health, and environment are integral to effective early management. In today's marketplace, optimal process and product design must consider employee and customer safety, and environmental protection and sustainability. Ultimately, effective early management enables an organization to introduce new products and processes rapidly, efficiently, and safely, and with minimal losses. The effective integration of the SHE and EM provides powerful tools for achieving zero accidents, zero incidents, and zero losses in the design and installation of new products, processes, and equipment.

5.6 Hazard and Operability Reviews

Many organizations aim to ensure process safety via compliance with codes and regulations. This approach is adequate when the applicable codes effectively anticipate and address all potential hazards and accident scenarios. Unfortunately, in many situations the codes only consider hazards and accidents that have arisen in the past. What is needed is a risk management approach that systematically identifies all potential hazards, evaluates them, and provides recommendations for improved process design and operation.

A key goal of early management is to design new products, processes, and equipment that are as safe and environmentally sound as possible, with the ultimate goal of achieving intrinsically safe and sustainable design. As stated by Ivars Peterson, "Success depends on an awareness of all possible failure modes, and whenever a designer is either ignorant of, or uninterested in, or disciplined to think in terms of failure, he can inadvertently invite it."[5]

To achieve optimal safety and sustainable design it is important to utilize systematic analysis techniques that tap into the wealth of expertise that exists throughout the organization. Hazard and operability reviews, known as HAZOPs,[6] are a type of structured review that enables a multidisciplinary team of experts to optimize the safety and functioning of equipment and processes. HAZOPs use specific guidewords and parameters to identify the causes and consequences of deviations from design intent. Table 5.1 outlines the common guidewords, parameters, and resulting deviations that are analyzed during a HAZOP study, and Table 5.2 provides a typical HAZOP worksheet for recording findings and recommendations.

5.7 FMEA and Start-Up Checklists

Failure mode effect analysis (FMEA) is an inductive, team-based methodology for identifying potential problems associated with a process, product, or service by investigating the potential failure modes for each element in the system. Originally utilized by the military and aerospace organizations for identifying and resolving operational, reliability, and safety issues, FMEA is currently utilized to address a wide range of process, product, and service problems.

FMEA can be used by the EM and SHE pillar teams during the design phase of a new system to identify ways in which a product, process, or service can fail to meet both customer and safety requirements. As indicated in Table 5.3, FMEA enables an organization to determine the failure modes of systems, the root causes of these failures, the effects of the failures, and potential corrective and mitigating actions. As part of the FMEA process the criticality or relative risk of each failure can be assessed by considering three factors: (1) the severity of the consequence, (2) the probability of its occurrence, and (3) the probability of the failure being detected. A failure's risk priority number (RPN) is determined by multiplying these three factors.[7,8]

FMEA provides a systematic approach for identifying problems and failure modes during early design stages. Start-up checklists and operational readiness reviews (ORR), on the other hand, can be used as a final check of a system's operability and safety prior to launch. These final SHE reviews

Table 5.1 Typical HAZOP Deviations Outlined in a Parameter and Guide Word Matrix

Parameter/ Guide Word	More	Less	No, None	Reverse	As well as	Part of	Other than
Flow	high flow	low or reduced flow	no flow	reverse flow	deviating concentration	leak or contamination	deviating or wrong material
Pressure	high pressure	low pressure	vacuum		delta-p		explosion
Temperature	high temperature	low temperature					
Level	high level	low level	no level, empty		different level		
Composition	more of one component	less of one component	no material present		contamination		new unexpected material
Concentration	high concentration	low concentration					
Corrosion	greater rate than expected						scaling, accumulation
Time	too long/too late	too short/too soon	sequence step skipped	backwards	missing actions	extra actions	wrong time
Agitation	fast mixing	slow mixing	no mixing				
Reaction	fast reaction/ runaway reaction	slow reaction	no reaction				unwanted reaction
Start-up/ Shut-down	too fast	too slow			actions missed		wrong recipe
Draining/ Venting	too long	too short	none		deviating pressure	wrong timing	

Table 5.2 Hazards and Operability Review (HAZOP) Worksheet

Guide Word	Parameter	Deviation	Cause	Consequence	Safeguard	Recommendation	Action

Table 5.3   FMEA Worksheet

Item Description	Function	Potential Failure Mode	Cause of Failure	Effect of Failure	Critical Rank/RPN[a] (severity × occurrence × detection)	Corrective Action

[a] RPN = Risk Priority Number = (Severity) × (Probability of Occurrence) × (Probability of Detection).

should be conducted during the commissioning stage or just prior to formal start-up of the project or process and transfer of the equipment to the operating area. A standardized checklist is used by a multidisciplinary early management team to prompt the thinking of the team members to check and ensure the proper functioning of all critical operational and safety items. A thorough operational readiness review is essential to ensuring a safe and vertical start-up.

5.8 30–60–90 Day Reviews

A key tool for ensuring effective early management and successful management of change is to conduct operability, safety, and sustainability reviews 30, 60, and 90 days after each change is implemented. These post-project safety reviews should be conducted to determine whether there are any remaining outstanding SHE issues and residual risks. In addition, these successive reviews reveal whether a change has been effective and whether there are any unforeseen operability, safety, or sustainability impacts and consequences.

Endnotes

1. Kletz, Trevor (1999). *Hazop & Hazan: Identifying and Assessing Process Industry Hazards*, Fourth Edition. Taylor & Francis. Philadelphia.
2. Nakano, Kinjiro (2003). *Planned Maintenance Keikaku Hozen: Comprehensive Approach to Zero Breakdowns*. Japan Institute of Plant Maintenance, Tokyo.
3. Shirosi, Kunio (1992). *TPM for Workshop Leaders*. Productivity Press, Portland, OR.
4. Gawande, Atul (2009). *The Checklist Manifesto: How to Get Things Right*. Metropolitan, New York.
5. Peterson, Ivars (1996). *Fatal Defect*. Random House, New York, p. 111. In Klentz, Trevor. 1999. *Hazop and Hazan*. Institution of Chemical Engineers. Rugby, UK, p. xi.
6. Royal Society of Chemistry (2010). "Environment, Health and Safety Committee Note on: Hazards and Operability Studies (HAZOP)." Retrieved on February 5, 2010 from http://www.rsc.org/images/HAZOPs_V2_190707_tcm18-95646.pdf
7. FMEA-FMECA.com (2010). "Your Guide to FMEA—FMECA Information." Retrieved on March 7, 2010 from http://www.fmea-fmeca.com/index.html
8. McDermott, Robin, R. Mikulak, and M. Beauregard (1996). *The Basics of FMEA*. Productivity Press, Portland, OR.

Chapter 6

Focused Improvement, Training and Education, and SHE

Capabilities can be extended indefinitely when everyone begins to think.

Taiichi Ohno[1]

6.1 Rationale behind Integrating Focused Improvement and SHE

By integrating its focused improvement (FI) and SHE processes an organization can more effectively identify, address, and eliminate the root causes of accidents and incidents and thereby achieve its vision of safety and manufacturing excellence: zero accidents, zero incidents, and zero losses. The focused improvement pillar provides powerful tools and techniques for achieving continual improvement (*Kobetsu Kaizen*) through formal, small-team problem solving that can be effectively utilized in the SHE area. Accidents can be eliminated and safety and sustainability performance can be improved by applying FI methodologies such as: 5-Why root cause analysis, Pareto charting, fishbone diagrams, process flow, PDCA, and CAPDo to safety, health, and the environment. By tapping into the collective creativity and intelligence of all employees the FI process enables an orga-nization to identify and implement new solutions to difficult and often long-standing safety, health, and environmental challenges.

6.2 SHE *Kaizens*

Safety, health, and environment should be managed like all other Lean or TPM pillars, based upon loss-tree data. Accidents and environmental inci-dents are reactive, or after-the-fact, loss data that can be used to identify areas that require SHE focused improvement. More importantly, information on near misses, SHE F-tags, and behavioral observations can be considered valuable pre-loss data or jewels that can be mined to bring real value to the organization. Pre-loss data enable one to proactively identify improvements that are needed to prevent accidents, incidents, and losses before they occur. SHE F-tags provide data on unsafe conditions in need of improvement, whereas formal safety observations provide information on at-risk behaviors that can be addressed via *kaizens*.

Formal improvement or *kaizen* teams should be established to tackle safety, health, and environmental issues that involve significant and chal-lenging losses, and can best be addressed by tapping into the collec-tive expertise of a group. SHE *kaizen* teams are typically established to address chronic SHE losses that have complex and multiple causes. Each organization should develop a *kaizen* register, as shown in Table 6.1, which lists the potential improvement projects covering all pillars that can be pursued at the site. The example *kaizen* register lists the safety, health, and environmental improvement projects that should be overseen by the SHE pillar team. The register includes the name or theme of the *kaizen*, a description of the project scope, an estimate of the losses that can be eliminated by the project, the estimated cost of the project, and the projected net savings from the project. Projects are added to the reg-ister by the organization's FI pillar leader based upon site loss-tree data. Organizations committed to achieving safety, health, and environmental excellence should ensure that some SHE improvement projects are always included on the site's overall *kaizen* register. A key role of the site's SHE pillar leader and team is to identify safety, health, and environmental improvement projects and to advocate for their implementation by the organization. The site's Lean or continuous improvement leadership group reviews projects on the register and approves and resources the projects.

Table 6.1 Example of a Kaizen Register

Area	Pillar	Kaizen Topic/Theme	Scope	Losses	Kaizen Cost ($)	Kaizen Net Savings ($)	Time	Resp.
Processing Safety	SHE	Elimination of failure to properly lockout equipment	All lines	10 LTAs, $1,000,000	$50,000	$950,000	Q4 2010	B. Safe
Packaging Health	SHE	Reduction in noise exposure	Lines 1–10	20 hearing loss cases, $600,000	$150,000	$450,000	Q1 2011	T. Bill
Production Environment	SHE	Reduction in site effluent BOD	Washouts	2 fines, 1 ton product, $800,000	$200,000	$600,000	Q2 2011	I. Green

	Idea Phase	Feasibility Phase	Capability Phase	Launch & Implement Phase		Evaluate Phase
# FI Projects	23	16	13	5	5	3
$ Value of Projects	5,500,000	4,000,000	3,500,000	1,500,000	1,000,000	1,000,000

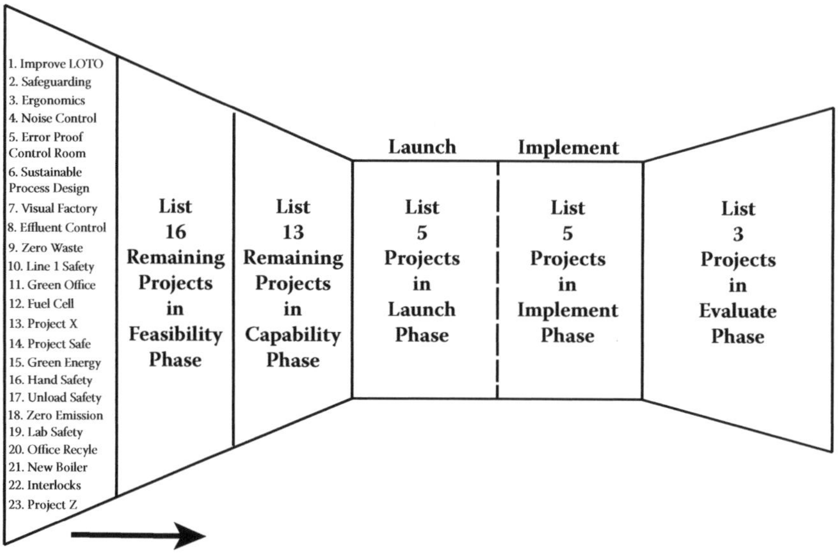

Figure 6.1   The SHE focused improvement funnel.

As shown in Figure 6.1, a focused improvement project funnel can be used to track the status of the organization's *kaizens* from idea generation to completion, the number of *kaizens*, and the total value of *kaizens*. A formal charter should be developed for each approved *kaizen* outlining the scope of the project, its implementation timeline, and expected deliverables.

The formal problem-solving process utilized to make focused improvements in the SHE area should be the same as that used by the other Lean or TPM pillars. Many organizations utilize the plan–do–check–act (PDCA) cycle or the similar check–act–plan–do (CAPDo) process as standardized approaches for implementing workplace *kaizens*. Commonly, the progress of implementing a SHE improvement is documented and communicated to all employees by means of a "*kaizen* story"[2] outlined on an activity board. A typical step-by-step problem-solving process utilized in SHE *kaizens* is as follows:[3]

Check Phase

Step 1. Identify the SHE Losses: Collect and classify the organization's safety, health, and environmental losses.

Step 2. Select the SHE Improvement Topic: Stratify the SHE losses utilizing tools such as Pareto charting, and then select the appropriate improvement topic based upon the data.

Step 3. Understand the Current and Ideal Situation: Go to the *gemba* or workplace in order to investigate and understand first hand the facts surrounding the specific SHE loss under study. Understand the principles of operation of all equipment involved in the incident and develop a working knowledge of applicable procedures and process flows. Contrast the current situation with the safety ideal in order to understand how the incident deviated from the ideal situation, and to define the ultimate goal of the SHE improvement project.

Step 4. Establish the *Kaizen* Objectives and Project Plan: Clearly define the scope, objectives, and time line of the SHE improvement project.

Analyze Phase

Step 5. Analyze Causes: Utilize various analytical tools such as: 5W1H, 5-Why root cause analysis, fishbone diagrams, A–B–C analysis, and process flow analysis to identify potential root causes of the incident. Test theories and utilize models and prototypes to verify the root causes.

Plan Phase

Step 6. Plan Countermeasures and Improvements: Draft alternative improvement proposals and compare their costs and benefits. Select the best options and develop a detailed implementation plan outlining the what, when, how, and who of each action.

Do and Review Phase

Step 7. Implement Countermeasures and Improvements: Implement each agreed SHE improvement action. Practice early management by employing appropriate test and acceptance methods, and by providing procedures and instructions for new operations.

Step 8. Verify the Results: Confirm the effectiveness of the SHE improvements and determine whether the agreed project targets have been met. If targets have not been achieved, return to Step 5 and analyze the causes again.

Step 9. Standardize the Improvement and Consolidate the Gains: Standardize and institutionalize improvements that have been demonstrated to be effective by developing written procedures, standards, and training. Replicate the improvements across the organization for similar equipment and situations.

Step 10. Formulate Future Plans: Identify additional improvements and *kaizen* projects that are needed to get the organization closer to realizing its safety, health, and environmental goals.

6.3 5S for SHE/6S: Condition-Based *Kaizens*

In most organizations many safety inspections are conducted and long lists of substandard workplace conditions are developed for someone else, typically maintenance, to correct. Because no one ever conducts a systematic root cause analysis on why the unsafe conditions existed in the first place, the same risky physical conditions continue to arise. Without identifying and addressing the root causes of substandard workplace conditions the organization is trapped in an endless cycle of repeated repairs of recurring unsafe conditions.

As previously mentioned, SHE F-tags are pre-loss data on substandard workplace conditions that can be used to proactively improve the work environment before someone is injured or the environment is harmed. Organizations can implement a formal *kaizen* process, known as 5S for SHE or 6S, to systematically analyze and address site SHE F-tag data. Figure 6.2 shows how 6S and the CAPDo problem-solving process can be used to systematically identify and remove at-risk conditions from the workplace. The first phase of the 6S process, the check phase, involves proactively seeing safety by detecting equipment and workplace problems before they become accidents. The *kaizen* team performs SHE audits of the work area to identify equipment abnormalities and substandard workplace conditions. By completing SHE F-tags, taking digital photos of unsafe conditions, and creating a visual map of workplace hazards the team creates a picture of the current state of SHE in their work area and documents baseline safety, health, and environmental conditions.

Whereas the check phase is about proactively seeing safety conditions, the analyze phase focuses on thinking deeply about workplace safety conditions. During this phase, the *kaizen* team studies and scrutinizes the SHE F-tag data and develops a Pareto chart or ranked histogram highlighting the most common unsafe conditions. The *kaizen* team then conducts an in-depth root cause analysis of the most common unsafe condition.

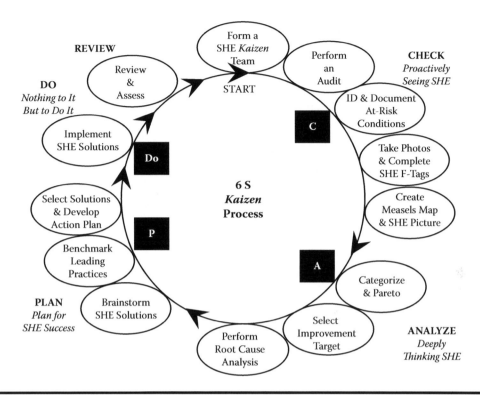

Figure 6.2   6S focused improvement *kaizen* process.

Once the root cause of an unsafe condition is identified the *kaizen* team can plan and implement appropriate corrective and preventive actions to remove the hazard from the workplace permanently. In the plan phase the team benchmarks current practices against leading SHE practices and brainstorms possible solutions in order to develop an action plan that is targeted and effective. The best solutions are identified and an action plan is developed that clearly delineates the specific measures to be taken, the timing of implementation, and associated responsibilities. In the do phase, the team facilitates implementation of the agreed corrective actions and countermeasures. In this way, the underlying root causes of unsafe conditions are systematically identified and removed from the workplace, and lasting safety and environmental improvement is achieved.

6.4 People-Centered Safety: Behavior-Based *Kaizens*

All organizations should implement a formal system for identifying and removing unsafe behavior from the workplace. Traditional behavior-based safety (BBS) processes provide a mechanism for eliminating risky behavior and promoting safe work practices. Through formal workplace safety observation and feedback processes, traditional BBS implementations provide employees with insights on how to work safer, and the organization with data on the types and frequency of unsafe behavior. Ideally, this safety observation data can be utilized to identify those safety work practices and behaviors that would benefit from formal focused improvement. Unfortunately, many BBS processes function as isolated activities separate from the overall management system, and outside the organization's continuous improvement process. As a result, many BBS processes do not effectively utilize BBS observation data for problem solving and implementing safety improvements.

People Centered Safety is a comprehensive BBS process that is integrated into an organization's Lean process, and therefore, is naturally linked with the Focused Improvement Pillar. People Centered Safety recognizes that successful safety programs care about people, and therefore, focuses not on organizational safety statistics and things, but on the individual. People Centered Safety realizes that safety excellence is not achieved by blaming accidents on employee behavior, but rather by creating a workplace culture that values safe work practices and integrates safety into the organization's way of working. People Centered Safety involves a series of improvement initiatives and activities designed to provide each employee with the insights, motivation, knowledge, skills, and ability to work safely at all times. People Centered Safety aims to establish an empowered organizational safety culture where employees achieve a state of autonomous safety—a state where employees take ownership of their own safety and the safety of their fellow employees. Employees take charge of their own risky behavior and are able to work without injury, even in the presence of hazards. In addition, employees are able to work safely in teams, are comfortable providing safety support to their colleagues, and are empowered to make safety improvements and conduct safety *kaizens*.

Similar to other BBS processes, a key element of People Centered Safety is a proactive, employee-led safety process that involves safety observation, feedback, and problem solving. People Centered Safety, however, differs from traditional BBS processes in that it integrates BBS into the Lean process and leverages Lean tools and problem-solving methodologies to remove barriers to safe work, and to promote safe work practices and behaviors. In People Centered Safety, employee workplace observation data is collected and managed like any other Lean or TPM loss data. The safety observation data is analyzed and charted in the facility like other loss data. For example, Pareto analysis is utilized to determine and communicate the most common at-risk behaviors. As part of the organization's overall Lean focused improvement process, safety improvement or *kaizen* teams are chartered to identify the root causes of the most common at-risk behaviors. Using 5-Why root cause analysis, antecedent-behavior-consequence analysis, and other Lean tools the root causes of the risky behavior are identified. Using the organization's standard problem-solving approaches, such as PDCA or CAPDo, the safety *kaizen* team evaluates potential corrective and preventative actions, and implements appropriate solutions to remove the barriers to safe behavior, and to facilitate safe work practices. By seamlessly integrating BBS into the Lean process, the organization ensures that the BBS *kaizen* process is adequately resourced and sustainable, and thereby better able to realize the goal of autonomous safety, and the vision of triple zero (zero accidents, zero incidents, and zero losses).

6.5 Integrating SHE into the Training and Education Pillar

The vision of the training and education (T&E) pillar is closely linked to the goals of the SHE pillar. The vision of the T&E pillar is to implement an effective training and education process that not only provides all employees with the knowledge and skills required to safely and effectively perform their jobs, but also provides them with the abilities to continuously improve company operations. The T&E pillar aims to unleash employee potential and creativity and thereby add value to the business.

An effective training and education process promotes a "safety comes first" culture, and provides powerful tools for achieving triple zero: zero accidents, zero incidents, and zero losses. Via the effective implementation of training programs, the verification of understanding, and the transfer of learning by means of one-point lessons (OPLs), the T&E pillar supports and promotes SHE excellence.

6.6 *Shu Ha Ri*

Shu Ha Ri are three Japanese terms that describe the progression of student learning from beginner to master. This traditional learning approach is used widely in Japan in martial arts academies, schools, and even industry. The *Shu Ha Ri* training approach is commonly utilized in Lean and TPM implementations because it is particularly useful for the development of basic skills, and in the multiskilling of workers. In the *Shu Ha Ri* approach the student learns via the repetition of basic skills and gradually develops mastery as the individual skills are integrated into an effective whole. Under the watchful guidance of a master, the learner progresses and the basic truths of the field are slowly revealed. Although this Japanese instructional and learning method is fundamentally different from the questioning approach of the Socratic method commonly used in Western societies, it provides some useful insights that can be used to improve workplace safety training.[4]

The first step, *Shu*, means to follow or obey. In the *Shu* phase the student follows and adheres to the teachings of the master, or *sensei*, learning the foundations of the discipline and developing technical understanding and competence. In the *Shu* phase instruction focuses on learning the rules and fundamentals by repetition and rote. These *Shu* learning concepts can be effectively applied to the safety training of new employees. During the *Shu* phase of safety training the focus is on ensuring that all new employees understand the fundamental principles of safe operation of their machines before working with the equipment. Repeated safety training and hands-on drills reinforce the basics of safe machine operation. Training on safe machine operation cannot be rushed, and students cannot advance to the next level of training until basic safety is mastered. In addition, new employees are paired with an experienced master operator who mentors and guides the novice employee, creating a structured and supportive learning environment. The guidance and nurturing provided by the master craftsman or operator enables the apprentice employee to develop the skills and thought processes required to work safely and efficiently.

The second phase, *Ha*, means to detach or break through. In this step, the students begin to apply on their own the principles that they have been learning. In addition, the "student must reflect on the meaning and purpose of everything that she or he has learned and thus come to a deeper understanding."[5] In other words, in the *Ha* phase, manufacturing students gain a deeper understanding of their equipment, its principles of operation, and how to ensure its safe functioning. Safety understanding is inculcated and embedded in the apprentices by having the *sensei* (instructor) give them individual safety problems to solve. The safety student is challenged to apply the principles of safe machine operation to real-life situations involving nonstandard situations and upset conditions. In this way, the apprentice develops deep, practical, and functional knowledge of how to ensure safe machine operation in standard and upset situations.

The third step, *Ri*, means to leave or transcend. In this phase, the student becomes a master and is able to apply what he or she has learned in new and novel ways. The manufacturing student possesses a deep understanding of the equipment and its safety. At this stage, the student's knowledge is so complete that he or she is able to teach others. Safety learning is reinforced by having the advanced student assist in training others.

In some organizations training charts and documentation are coded to indicate the level of mastery achieved by an employee. Workers who have completed the *Shu* stage are simply considered trained but not fully qualified. Those that have successfully completed the *Ha* phase are considered qualified, and those who complete the *Ri* level are considered master operators who are able to teach others.

6.7 One-Point Lessons (OPLs)

In Lean and TPM implementations, one-point lessons are a key mechanism for conducting training and for communicating important operational and safety information. As the name implies, one-point lessons should focus on communicating one basic point (or at most a few) in an easy to understand format. Ideally, the lesson should be communicated in a visual way using diagrams or photographs. Organizations typically have several types of OPLs:

1. *Basic Knowledge OPLs:* For communicating information on basic machine design, principles of operation, and operating and maintenance procedures.
2. *Improvement Case OPLs:* For transmitting information on improvements and changes made to machine design, operating procedures, and maintenance.
3. *Problem Case:* Provides information on how to prevent an actual equipment problem or product defect.
4. *SHE OPLs:* For communicating safety, health, and environmental concerns; SHE procedures; prevention information; SHE improvements; and learning from accidents and incidents.

To ensure that all affected employees receive and understand the information conveyed in a one-point lesson, comprehension is verified through a test and through periodic follow-up on the plant floor. All affected employees are required to indicate their understanding of an OPL by signing or initialing the bottom of the form. Safety OPLs should be reviewed with all affected employees at least once a year.

6.8 SHE Training Plan and *Kaizen* Gym

Organizations should have systems in place to identify employee safety, health, and environmental training needs and requirements. Training needs should be based on specific job duties or tasks, potential hazards, and company standards, as well as legal requirements from OSHA, EPA, DOT, and local governmental agencies. As indicated in Tables 6.2 and 6.3, SHE training plans outline the type and frequency of training required for each role. The success and effectiveness of the SHE training plan should be assessed annually, and the plan should be modified and updated accordingly.

To make operational and safety, health, and environmental training practical and attention-grabbing, many organizations establish a separate training area called a "*kaizen* gym" or "safety gym" where hands-on instructional activities are conducted. The *kaizen* gym strengthens employee knowledge, understanding, and skills by engaging them in learning exercises involving real equipment, machine mock-ups, simulations, and demonstrations. The *kaizen* gym also includes examples of equipment defects and abnormalities that can lead to accidents and losses. This type of training increases and reinforces learning because it uses equipment the employees work with and involves them in real work-life situations.

In the SHE area, new employees can be involved in SHE 101 activities that engage them in hands-on activities with equipment and models that teach them about the principles of safe machine operation. Practical exercises and simulations are conducted involving machine de-energization and lockout/ tagout. New employees visit a museum of damaged parts and equipment to understand the root causes of equipment failures and safety incidents.

In Lean and in SHE the purpose of training is to increase employees' knowledge and skill so that they are able to perform their tasks and jobs in a safer and more effective fashion. Training without improvement in safety and job performance is ultimately a waste of organizational resources. Thus, the Lean approach to SHE and operational training focuses on practical, step-by-step training that steadily increases employee skills and capabilities. Typically, workers are ranked and categorized into five skill levels, as follows:[6]

Table 6.2 Safety and Health Training Needs Assessment

Occupational Safety and Health Training Subject	SH Manager[a]	Dept. Manager	Supervisor	Waste Handler	Maintenance	Hazmat Team	First Aid Team	Production Operator	Warehouse/ DC	Lab
Occupational Health and Safety Orientation	I	I	I	I	I	I	I	I	I	I
Annual Occupational Health and Safety Awareness Training	A	A	A	A	A	A	A	A	A	A
Detailed SHMS Review	A	A								
Hazard Communication	A	A	A	A	A	A	A	A	A	I, U
Basic Emergency and Evacuation Procedures	A	A	A	A	A	A	A	A	A	I
Hazardous Material Spill Response	I, P		A		A	A			A	
First Aid						P				
CPR/AED						A				

(Continued)

Table 6.2 Safety and Health Training Needs Assessment (Continued)

Occupational Safety and Health Training Subject	SH Manager[a]	Dept. Manager	Supervisor	Waste Handler	Maintenance	Hazmat Team	First Aid Team	Production Operator	Warehouse/ DC	Lab
Blood Borne Pathogens Exposure Control						A				
Noise Hazards and Hearing Conservation	A	A	A	A	A	A	A	A	A	I
HAZWOPER	I, P				A					
Asbestos	I			A						
Chemical Hygiene Plan/Lab Safety	I		A						A	
Comprehensive OH Regulations Course	I, P									
Confined Space Entry	I			A		I				
Ammonia	I	I	I	I	I	I	I	I	I	I
PPE Requirements	I, C	I, C	I, C	I, C	I, C	I, C	I, C	I, C	I, C	I, C
Ergonomics/Lifting	I, P	I, P	I, P	I, P			I, P	I, P	I, P	I, P

Occupational Safety and Health Training Subject	SH Manager[a]	Dept. Manager	Supervisor	Waste Handler	Maintenance	Haz Mat Team	First Aid Team	Production Operator	Warehouse/DC	Lab
Hot Work Permits	I	I		I, P						
Respiratory Protection	I, P	I		A	A					A
LOTO	I, P	I		A	A					
Work at Heights	I, P	I		A			A			
Pipe Breaking	I, P	I		A						
Process Safety Management(PSM)	I	I		P, C			P, C			
Forklift Operation	I	I						I, P		
Electrical Safety Related Work Practices	I			I						

a I = Initial, A = Annual, C = When Changes Occur, P = Periodic (every three years).

Table 6.3 Environmental Training Needs Assessment

Environmental Training Subject	Position										
	Environmental Coordinator[a]	Area Manager	Supervisor	Waste Handler	Maintenance	Hazmat Team	Lab	Production Operator	Warehouse/DC	Engineers	Office Staff
1. Environmental Orientation	I	I	I	I	I	I	I	I	I	I	I
2. Annual Environmental Awareness Training	I	A	A	A	A	A	A	A	A	A	A
3. Detailed EMS Review	A	A	A								
4. Basic Emergency Procedures	A	A	A	A	A	A	A	A	A	A	A
5. HAZWOPER	I					I, A					
6. PSM/RMP	A	A	A		A	A	A	A	A	A	
7. RCRA Waste Disposal	A		A	A							
8. Wastewater, Storm Water	A	A	A								
9. Air Emissions Inventory	A	A	A								
10. Environmental Regulation Course	I, P										

11. Environmental SOPs			A			A	A	A
12. Noise	A	A	A	A	A		A	A
13. Waste Minimization and Recycling	P	P	P	P	P	P	P	P
14. Sustainability	I	A	A	A	A	A	A	A

[a] I = Initial, A = Annual, P = Periodic (every three years).

Level 0: Lack of both knowledge and skill (needs both theoretical and practical training)

Level 1: Possesses only theoretical knowledge (needs practical training)

Level 2: Has knowledge but can only partially perform task (needs more practical training)

Level 3: Has knowledge and has demonstrated mastery of skill, but cannot teach

Level 4: Has mastered knowledge and skill, and can teach

SHE and operational training should be structured to advance employees through each successive skill level. For critical safety skills related to their jobs employees must attain Level 3 or Level 4. The highest level of skill attainment, Level 4, is that of a teacher or *sensei* where one has reached a high level of mastery and is confident and capable of instructing others. Ultimately, knowing the safe job procedure is not enough; the employee must be able to demonstrate that he or she has mastered the safety skill and can perform it properly. The Lean objective is to develop employees who are 100% competent in their tasks and in safety.

Endnotes

1. Ohno, Taiichi (1986). *Kanban Just-in-Time at Toyota*. In Japan Management Association. Productivity Press, Portland OR, p. 10.
2. Imai, Masaaki (1997). *Gemba Kaizen. A Commonsense, Low-Cost Approach to Management*. McGraw-Hill, New York, p. 58.
3. Suzuki, Tokutaro (1994). *TPM in Process Industries*. Productivity Press, New York, pp. 52–59.
4. Digenti, Dori (1996). "Zen learning: A new approach to creating multi-skilled workers." Center for International Studies, Massachusetts Institute of Technology Japan Program. Boston, MA. Retrieved on April 14, 2010 from: http://dspace.mit.edu/bitstream/handle/1721.1/7574/JP-WP-96-29.pdf?sequence=1
5. Fox, Ron (1995). "The Aikido FAQ: Shu Ha Ri." *The Iaido Newsletter*. 7,2: 1. Retrieved on April 28, 2010 from: http://www.aikidofaq.com/essays/tin/shuhari.html
6. Shirosi, Kunio (1996). *Total Productive Maintenance: New Implementation Program in Fabrication and Assembly Industries*. Japan Institute of plant Maintenance (JIPM), pp. 424–430.

Chapter 7

SHE Pillar Activities

The road must be run safe first, and fast afterward.

Rule Book of the New York and Erie Railroad[1]

7.1 SHE and Lean

Despite the limited discussion of safety, health, and environmental (SHE) issues in the Lean literature, it is obvious that accidents and environmental incidents are a form of waste that must be managed, reduced, and ultimately eliminated. Failure to integrate SHE into the Lean process can give rise to a business culture that endorses risk-taking, resulting in changes that are inadequately examined and, at times, even reckless. This failure to adequately manage the safety and environmental impacts of change presents unnecessary risk to the business, which can result in significant human and financial losses. Recent accidents such as the explosion of the BP Deepwater Horizon oil rig in the Gulf of Mexico and the resulting oil spill clearly demonstrate the high human, environmental, and financial costs of failing to manage SHE risks properly. Changes made in the name of speed, increased productivity, and lower cost that compromise safety and the environment are not improvements at all, but rather reckless betting with the lives and livelihood of others that has no place in today's Lean workplace. In fact, continual improvement in an organization's safety record is considered such a key indicator of the success of a Lean/TPM implementation that reducing accidents is a mandatory requirement for winning the Japanese PM prize.

Lean methods profoundly change the workplace and the way work is conducted. To ensure that workplace changes are both safe and sound it is essential that an organization's safety, health, and environmental systems adapt and respond to these changes. Early in Lean implementation an organization's SHE systems should be rethought and re-engineered. Lean offers an opportunity to integrate safety, health, and environment into the organization's way of working, rather than managing them as separate issues. The SHE pillar plays a key role in this integration process by ensuring that

1. The changes brought about by the Lean transformation are adequately reviewed from a safety, health, and environmental perspective.
2. All of the Lean/TPM pillars incorporate safety, health, and environment into their master plans and activities.
3. The Lean methodology and associated tools are leveraged by the SHE pillar to reduce risks, improve SHE performance, and advance the business.
4. Safety, health, and environmental risks and opportunities are factored into the business's decision-making process and its operations.[2,3]

7.2 SHE Pillar Master Plan and SHE Management Systems

In a typical Lean/TPM implementation, the workplace often undergoes challenging and wrenching transformations. By carefully formulating and implementing a comprehensive safety, health, and environmental pillar master plan, an organization is better able to manage this change in a safe and sustainable fashion. Developing a SHE master plan enables an organization to develop a clear roadmap for realizing its vision of safety, health, and environmental excellence.

The SHE pillar master plan is a Gantt chart, developed by the site SHE pillar team and endorsed by the organization's leadership, that provides the detailed schedule and responsibilities for implementing specific SHE initiatives and improvement activities. Figure 7.1 provides an example of a site SHE pillar master plan that shows how safety, health, and environment are integrated into the other pillars. The master plan includes activities that the SHE pillar conducts in support of the other pillars, as well as initiatives that the SHE pillar champions and leads with the support of SHE professionals.

SHE Pillar Master Plan

#	Tasks	Owner	Plan Start	Plan Finish	YEAR 1												YEAR 2											
					J	F	M	A	M	J	J	A	S	O	N	D	J	F	M	A	M	J	J	A	S	O	N	D
1	**SHE Support in Preparation Phase**																											
1.1	SHE Climate Analysis Support																											
1.2	SHE Leadership Training																											
1.3	Leadership 7 Cs of Safety																											
1.4	SHE Vision & Values																											
1.5	Zero Accident Education																											
1.6	Develop & Implement SHE Loss Tree																											
1.7	Implement SHE Governance																											
1.8	Propose SHE Targets																											
1.9	Draft SHE Balanced Scorecard																											
1.10	6 S Support																											
1.11	Support of LOTO/ZES																											
2	**SHE Support in AM Steps 1–3**																											
2.1	Danger Anticipation Training																											
2.2	SHE F-Tags																											
2.3	SHE OPLs																											
2.4	SHE 5-Why RCA																											
2.5	Support AM Elimination of Sources of Leaks, Spills,																											
2.6	Zero Leaks & Spills																											
2.7	SHE in Provisional AM Standards																											
2.8	SHE Visual Controls																											
2.9	Pointing & Naming Drills																											
2.10	SHE Loss Tree																											
3	**SHE Support in AM Steps 4–7**																											
4	**SHE Support of PM & EEM**																											
4.1	SHE Permit System Technical Support																											
4.2	Technical Support of Haz Ops & FMEA																											
5	**SHE Support of FI = SHE *Kaizens***																											
6	**SHE Pillar**																											
6.1	SHE Master Plan & SHE Mgmt System Design																											
6.2	Design of SHE Pillar Self-Assessment																											
6.3	SHE Charting																											

Figure 7.1   Example of a SHE pillar master plan.

In the Lean preparation phase the SHE pillar team is actively engaged in activities that build a strong and sound foundation for the organization's SHE management systems and the overall Lean initiative. These foundational SHE pillar activities include assessing the organization's safety climate and leading a series of workshops focusing on SHE leadership that aim to align management and employee beliefs with the company's vision of safety, health, and environmental excellence. Once the organization is aligned around a common vision of SHE excellence, the SHE pillar facilitates the development of a balanced scorecard and associated safety, health, and environmental targets. As employees organize and ready their work areas for the Lean implementation by applying 5S principles, the SHE pillar facilitates the expansion of 5S methodologies into the SHE area via 6S.

The SHE pillar team works closely with the pace-setting autonomous maintenance pillar to implement activities that support the rollout of each AM step. As each AM step introduces new knowledge, improvement tools, and problem-solving methodologies to the workforce, the SHE pillar leverages these freshly gained insights and skills to improve workplace safety, health, and environment. For example, as work areas tackle the initial steps of autonomous maintenance and implement F-tags, one-point lessons, 5-Why root cause analysis, and visual controls, the SHE pillar ensures that these tools are applied to the safety, health, and environmental area in order to reduce SHE losses and to improve SHE performance. In the later steps of autonomous maintenance, the SHE pillar facilitates the application of hands-on experiential training, enhanced visual controls, error-proofing, and standard work to address SHE issues.

The SHE pillar also collaborates with the other Lean/TPM pillars to ensure that their unique tool kits are applied to reducing safety, health, and environmental losses. For example, the focus improvement pillar's *kaizen* methodology is utilized to conduct systematic SHE improvement projects. The early management pillar's HAZOP and FMEA methodologies are employed to improve the SHE aspects of new equipment and product design. Early management's concepts of error-proofing and fail-safe design are applied to eliminate accidents, incidents, and SHE losses. In addition, the SHE pillar works closely with the training and education pillar to ensure that relevant safety, health, and environmental principles and practices are incorporated into all training programs. In the later stages of Lean, as the organization embarks on zero loss initiatives, the SHE pillar sponsors zero accident and zero environmental incident activities.

The SHE pillar plays a leading role in the design and assessment of the organization's safety, health, and environmental management system, and in ensuring the linkage of the management system with the overall SHE pillar plan. Regardless of which management system model an organization uses, the essential function of the SHE management system is to systematically identify and eliminate the hazards that are the root causes of workplace accidents, environmental incidents, and SHE losses. All organizations have limited resources; therefore, it is essential that the SHE management system and associated activities be focused on and driven by the safety, health, and environmental loss tree.

Many leading organizations utilize internationally accepted management system consensus standards such as ISO 14001[4] and OHSAS 18001[5] as the basis of their SHE management systems. Both the ISO 14000 (International Standards Organization Environmental Management Systems—Requirements with Guidance for Use) and the OHSAS 18001 (Occupational Health and Safety Management Systems—Requirements) management system models use a plan–do–check–act or PDCA cycle to ensure the continued effectiveness of the organization's SHE process. As shown in Figure 7.2, the 17 basic elements of an ISO-based SHE management system are

1. SHE Policy
2. Health and Safety Hazards and Environmental Aspects Identification
3. Legal and Other Requirements
4. Objectives and Targets
5. SHE Management Program(s)
6. Structure and Responsibility
7. Training, Awareness, and Competence
8. Communication
9. SHE Management System Documentation
10. Document Control
11. Operational Control
12. Emergency Preparedness and Response
13. Monitoring and Measurement
14. Nonconformance and Corrective and Preventive Action
15. Records
16. SHE Management System Audit
17. SHE Management Review

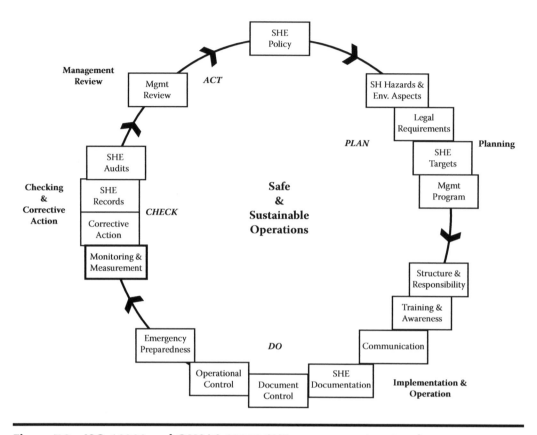

Figure 7.2   ISO 14000 and OHSAS 18001 SHE management system improvement model.

A SHE management system is not meant to be a voluminous set of manuals that sits on a bookshelf collecting dust. The SHE management system is meant to be a living set of protocols that organize the business's SHE process and define how the organization will identify, assess, manage, and ultimately control its significant safety, health, and environmental risks. The heart of an effective SHE management system is risk management. A successful management system is not reactive and compliance based, but rather is proactive and risk based. It employs leading indicators of both condition-based risk and behavior-based risk such as near misses, equipment F-tags, and behavioral observations to guide decisions and actions.

The management system is put into action and comes alive by integrating it with the Lean/TPM process via the implementation of proactive projects that identify and remove physical and behavioral risks from the workplace. To achieve this objective, some organizations establish a core SHE process that defines the routine safety, health, and environmental activities that must be conducted in order to meet the established SHE targets and the vision of SHE excellence. The core SHE process clearly delineates what SHE activities will be conducted, who is responsible for completing them, where they will be implemented, when they will be performed, and how and how frequently the activities will be carried out. In other words, it defines the 5Ws and 1H (what, who, where, when, why, and how) of the SHE activities. Table 7.1 outlines how an organization's SHE management system and its Lean/TPM SHE activities can be effectively linked and leveraged. The SHE management system is not a separate set of standards implemented in isolation, but instead it consists of vital procedures and activities that are seamlessly integrated into the organization's Lean process and ways of working.

To ensure that the SHE activities essential to safety success are implemented consistently and effectively, many leading organizations utilize checklists and visual controls to structure and direct work execution. In the *Checklist Manifesto*,[6] Atul Gawande provides convincing evidence that the use of a humble checklist can improve the safety and job performance of a wide range of tasks. Checklists, by providing a physical reminder that ensures that all required steps in a task are completed properly, can reduce the number of mistakes made by new and inexperienced workers. Checklists, however, are useful not only for the unskilled worker performing mundane tasks, but can also decrease the errors made by experienced and highly skilled workers performing complex tasks. Professional airplane pilots have used checklists effectively for performing preflight safety checks and for dealing with inflight emergency situations. Medical institutions have successfully utilized checklists to decrease physician errors and increase patient safety in operating rooms and emergency rooms. Because most jobs are not mistake-proof, and most people are not immune to making errors, no one should be too proud to use a checklist. A checklist is not a crutch but rather an essential tool for improving job performance and safety. The proper completion of Lean SHE activities is essential for safety, health, and environmental success; therefore, winning organizations utilize checklists widely to ensure the effective execution of SHE work. In successful organizations the SHE management system structures the overall risk management process, the core SHE process defines the "5Ws and 1H" of the SHE work that is necessary to achieve SHE success, and checklists are utilized to ensure that the SHE activities are completed properly.

Table 7.1   Linkage of SHE Management Systems and Lean SHE Activities

PDCA	No.	ISO 14001/OHSAS Element Title	Lean/TPM SHE Activities	
Policy	1	SHE Policy	SHE Vision and Values	
			Sustainability Policy and Principles	
Plan		Planning		
	2	Hazard and Identification	SHE Value Stream Mapping	FMEA
		Environmental Aspects Identification	SHE F-Tags	BBS Observations
			Hazard-Operability Reviews	Environmental Mapping
			SHE Loss Tree	Industrial Hygiene Mapping
	3	Legal and Other Requirements	SHE Standards	
	4	Objectives and Targets	SHE Targets	
			SHE Balanced Scorecard	
	5	SHE Management Programs	SHE Master Plan	
Do		Implementation and Operation		
	6	Structure and Responsibility	SHE Governance	
			SHE Pillar Team	
			Area SHE Teams	
	7	Training, Awareness, and Competence	SHE Training Master Plan	Zero Accident Education
			SHE Leadership Training	One-Point Lessons
			7Cs of Leadership	Danger Anticipation
			SHE *Kaizen* Gym	Experiential Training
			AM/SHE Training	AM/PM/SHE Partnership

8	Communication	TPM SHE Pillar Kickoff Event
		Pillar Team Meetings
		SHE Activity Boards
		SHE One-Point Lessons
9	SHE Management System Documentation	ISO and OHSAS SHE Mgmt System Manual
		Core SHE Process Manual
10	Document Control	5S in Administration
11	Operational Control	LOTO/ZES	Work Permit Systems
		5S and 6S	SHE Standard Work
		AM Standards	BBS/PC Safety
		Visual Controls	Error-Proofing
		Core SHE Process	Intrinsically Safe Design
		Zero Accident Initiatives	Zero Waste Initiatives
		Sustainability Initiatives	
12	Emergency Preparedness and Response	Modular, Checklist-Based, and Visual ER Plans
Check	Checking and Corrective Action	
13	Monitoring and Measurement	AM/SHE Inspections
		SHE Inspection Points
		SHE Checklists
		30–60–90 Day Reviews
		Behavioral Observations and Feedback (BBS OF)

(Continued)

Table 7.1 Linkage of SHE Management Systems and Lean SHE Activities (Continued)

PDCA	No.	ISO 14001/OHSAS Element Title	Lean/TPM SHE Activities
	14	Accidents, Incidents, Nonconformance, and Corrective and Preventive Action	5-Why Root Cause Analysis (RCA)
			One-Point Lessons (OPLs)
			SHE *Kaizens*
			SHE Control Charting
	15	Records	5S in Administration
	16	Environmental Management System Audit	SHE Climate Analysis
			ISO Validation Audits
			SHE Audits
Act	17	Management Review	SHE Pillar Self-Assessment
			SHE Management System Reviews

Finally, leading organizations design and implement their SHE management systems in a forward-thinking and proactive fashion focusing on

1. Preventing all SHE accidents, incidents, and losses
2. Continual improvement in safety, health, environmental, and sustainability performance
3. Leveraging SHE issues for competitive advantage
4. Promoting environmentally sound and sustainable actions that conserve resources and prevent degradation of the natural environment

A SHE management system based on these principles will be effective at reducing the business's SHE risks, but will also be responsive to meeting customer and market needs in the safety, health, environmental, and sustainability areas.[7]

7.3 SHE Pillar Self-Assessment

Lean, like any other continuous improvement initiative, requires periodic assessment to determine the degree of success and progress toward meeting established standards. Likewise, the SHE pillar should undertake regular self-assessments to gauge the organization's progress toward

1. Reducing safety, health, and environmental losses
2. Implementation of the SHE pillar master plan
3. Application of Lean methodologies and tools in the SHE area
4. Integration of sustainability principles and practices into the business

Typically, an initial baseline assessment of the organization's SHE process is undertaken during the Lean preparation phase in order to obtain an initial snapshot of the state of safety, health, and environment in the business. Subsequent pillar self-assessments are commonly conducted every six months to gauge the progress toward implementing the Lean TPM pillar process.

As the name implies, SHE pillar self-assessments are conducted by an internal team familiar with the site SHE process, but able to provide an accurate and objective evaluation of the functioning of the SHE pillar. The self-assessment is not a regulatory compliance audit, but an examination of the state of the SHE pillar. It consists of a series of standard questions

SHE Pillar Self Assessment

Element 2: Identification of Safety & Health Hazard & Environmental Aspects

2.1 Are area management and employee teams involved in periodic identification and evaluation of safety & health hazards associated with routine tasks?		Score
0	No formal safety & health hazard identification & evaluation process exists.	
1	Area management and employees are informed of task risk assessments performed by others.	
2	Area management and employees have identified and evaluated the safety & health hazards of some routine tasks.	
3	Area management and employees have identified and evaluated the safety & health hazards of most routine tasks but the assessments are not fully up to date.	
4	Area management and employees have identified and evaluated the safety & health hazards of most routine tasks and have kept them current. The documentation of task risk assessments is not uniform and OPLs have not been used to transmit key learning.	
5	*Area management and employee teams have been trained in performing risk assessments on routine tasks, and at least annually are conducting formal safety & health assessments. The task risk assessments have been document via Task Hazard Analysis (THA), or Job Safety Analysis (JSA). One Point Lessons (OPLs) are used to transmit key learning from risk assessments.*	
Assessor Findings & Comments:		
2.2 Does the site have a process for systematically identifying & evaluating at-risk or unsafe conditions in the workplace?		Score:
0	No process exists for identifying and evaluating at-risk conditions.	
1	A site safety committee conducts periodic area physical condition safety inspections.	
2	Area supervision conducts periodic inspections in which lists of unsafe conditions needing correction are developed. No root cause analysis of findings is conducted.	
3	Area teams conducts periodic inspections in which lists of unsafe conditions needing correction are developed.	
4	Area teams identify at-risk conditions using SHE F-tags. A summary of the SHE F-tags and key safety issues is posted.	
5	*Area teams systematically identify at-risk conditions using a formal SHE F-tag process. Measles or dot maps of F-tags are posted highlighting the locations of at risk condition. Via a 6S process all SHE F-tags are analyzed and the root causes of repeated unsafe conditions are identifies and addressed by Kaizen teams.*	
Assessor Findings & Comments:		
2.3 Does the site have a process for systematically identifying & evaluating at-risk or unsafe behavior in the workplace?		Score:
0	No process exists for identifying and evaluating at risk-behavior.	
1	Supervision conducts periodic observations of work related behavior and compliance with safe job procedures.	
2	Employees conduct formal safety observations & provide feedback. Observation data is not posted.	
3	Employees conduct formal safety observations & provide feedback. Observation data is summarized & posted	
4	Employees conduct regular safety observations & provide feedback. Safety observation data is posted and analyzed to identify root.	
5	*Employees conduct regular safety observations & provide feedback. Safety observation data is posted and analyzed to identify root. Kaizen teams are established to implement improvements that will promote safe work practices and eliminate behavioral risk.*	
Assessor Findings & Comments:		
2.4 Has the organization developed a SHE Loss tree and is the SHE loss tree used to identify and evaluate safety and health hazards in the workplace?		Score:
0	The site has not developed a SHE Loss Tree.	
1	The site has developed a basic SHE Loss Tree.	
2	The site has developed a basic SHE Loss Tree of lagging indicators. Past accident and incident data is used to identify and evaluate workplace risks.	
3	The site has developed a SHE Loss Tree with some leading and lagging indicators, some of which are used to identify and evaluate workplace risks.	
4	The site has developed a comprehensive SHE Loss Tree including of both leading and lagging indicators. However the loss tree data is infrequently analyzed.	
5	*The site has developed a comprehensive SHE Loss Tree including of both leading and lagging indicators. Data from the loss tree (near misses, at-risk conditions, at-risk behavior, workplace exposures...) are regularly analyzed to identify potential safety and health hazards.*	
Assessor Findings & Comments:		

Figure 7.3   Example of SHE pillar self-assessment scoring.

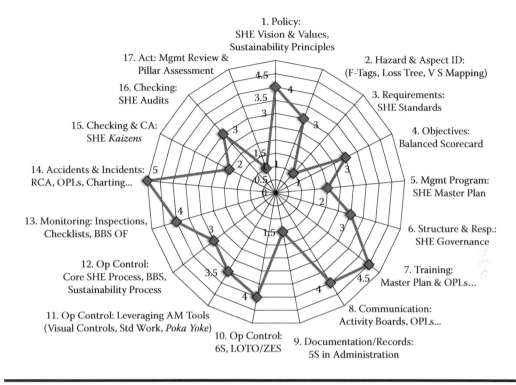

Figure 7.4   SHE pillar self-assessment radar chart.

related to the development, implementation, and functioning of the SHE pillar's methodologies and tools. As shown in Figure 7.3, a score from 0 to 5 is provided for each question assessing the relative strength of the facility's SHE pillar in each particular area. The evaluators use standardized definitions that describe full performance and the various levels of suboptimal performance related to each SHE pillar element to ensure that the ratings are applied accurately and consistently.

Commonly, a radar or spiderweb chart is used to visually depict the results of the self-assessment by graphing the average scores that the organization has obtained for the major areas of SHE pillar implementation. As shown in Figure 7.4, the self-assessment radar chart displays the organization's SHE pillar strengths and weaknesses, and the magnitude of the gaps from optimal performance. It is important to ensure that the SHE pillar self-assessment categories and questions are closely linked to the organization's SHE management system and its Lean/TPM process. It is only fair that the chosen assessment tool accurately evaluates the SHE pillar on those elements and activities that the organization mandates and believes will deliver

SHE success. The example self-assessment radar chart in Figure 7.4 contains 17 major categories linked to both the ISO-based SHE management system model and the Lean SHE pillar methodology. This linkage and integration of the organization's SHE management system and Lean process in the pillar self-assessment reinforces the principle that SHE is not a separate activity, but is integral to the operation of the business.

The most important aspect of the self-assessment process is the development of an action plan to close the gaps and eliminate the weaknesses that have been identified. As with any action plan, the SHE pillar action plan should clearly delineate what corrective and improvement actions will be taken, who is responsible for implementing them, and the dates by which they will be completed. The SHE pillar can complete a SWOT analysis as shown in Table 7.2 to identify the organization's SHE strengths, weaknesses, opportunities, and threats. As part of the SWOT analysis the SHE pillar team should identify the actions to be taken to

1. Reinforce and leverage the strengths of the SHE process.
2. Correct the weaknesses of the SHE management system.
3. Capitalize on the opportunities available to the business in the safety, health, environmental, and sustainability area.
4. Plan for and address the threats facing the business that could affect SHE.

7.4 SHE Control Charting

Safety, health, and environmental management within an organization is a process, and as a process its performance is subject to variation. As is any other process, the SHE process is subject to two types of variation that have different sources. The first type of variation, known as common cause variation, is inherent within the process due to its design and normal operation. Common cause variation, also known as normal variation, is consistent and predictable. As a result of common cause variation the safety performance of an organization whose safety process is under statistical control is not a straight line, but varies slightly within distinct limits defined by the process. In other words, even a safety process under control will experience slight changes in the number of accidents each month due to normal variation. Because common cause variation is inherent in the SHE process it is not easily addressed by shop floor personnel. To reduce common cause variation in a stable SHE process it is necessary for management to make fundamental

Table 7.2   Example of a SHE Pillar SWOT Analysis

Area	Strengths	Weaknesses	Opportunities	Threats
1. SHE Policy	S1.1 SHE policy is well communicated to all levels of the organization. *Action: Continue with annual SHE policy communication cascade.*	W1.1 Process for updating SHE policy is not well defined. *Action: Establish a formal procedure for annually updating the SHE policy that involves the SHE council and SHE pillar team.*	O1.1 SHE policy does not consider opportunities in the area of sustainability. *Action: Convene a leadership workshop to consider expansion of the SHE vision and values, and the SHE policy in the sustainability area.*	T1.1 Recent acquisition involves integration of a new business and new employees with different SHE vision and values. *Action: Conduct SHE vision and values training and zero-accident education for all new employees. Expand the SHE Council and SHE pillar membership to include reps. from the new business.*
	S1.2	W1.2	O1.2	T1.2
2. SHE Hazards and Aspects				
3. Legal Requirements				
4. Objectives and Targets				
5. SHE Mgmt. Program(s)				

changes in the SHE process that result in the process operating at a new and improved level. The second type of variation is called special cause variation and, as the name implies, occurs as a result of a unique change or upset in the SHE process. Special cause variation is inconsistent and unpredictable.[8]

The use of statistics and statistical process control (SPC) charts enables one to differentiate common cause variation in the SHE process from variation due to special cause. Figure 7.5 shows an SPC chart of an organization's recordable accident rate. The initial part of the chart depicts the safety process in control and the accident rate varying within the upper control limit (UCL) and lower control limit (LCL) due to normal or common cause variation. Because the control limits are plus and minus three standard deviations from the mean, over 99% of the data falls within the UCL and LCL. Based upon this and other statistical principles, several rules have been developed for interpreting control charts and determining whether a process is in control or out of control due to the presence of special cause. Rule no. 1 for control charts states that whenever any data point is outside the control limits special cause is present. This and other SPC charting rules are outlined by Mal Owen in his book, *SPC and Continuous Improvement*.[9]

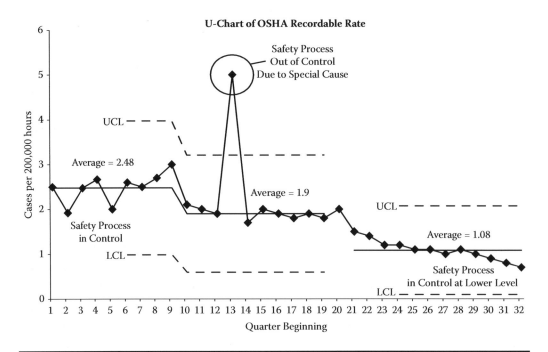

Figure 7.5 Example of an SPC chart (U-chart) of recordable accident rate.

The middle section of the SPC chart in Figure 7.5 shows the accident rate out of control and rising sharply with a data point above the UCL. Over 99% of the data are expected to fall within the process's control limits; therefore, this point is an outlier. According to SPC charting Rule no. 1, this spike in the recordable accident rate is due to special cause. At this point it is important for the organization to investigate and identify the special cause that has resulted in the accident increase, and to take corrective action to address it in order to bring the accident rate back to normal levels. The final part of the chart reveals that the organization has made fundamental improvements in its safety process that have reduced common cause variation and established new, lower control limits for the organization's recordable accident rate. Figure 7.5 demonstrates that SPC or control charting can aid organizations in analyzing their SHE process by providing clues to the source of process variation. In addition to this advantage, SPC charting of SHE data offers the following benefits:

1. Provides increased visibility and attention to facility SHE data and the functioning of the SHE process
2. Increases understanding of the organization's SHE process and its operation
3. Enables personnel to more rapidly identify significant adverse changes in the SHE process that require action from natural random noise in the process
4. Minimizes management tinkering and tampering with the SHE process due to normal variation inherent in the process
5. Provides feedback on the results of interventions and actions taken to address the SHE process
6. Enables improved decision making on the actions to be taken to address variation in SHE performance

If an organization's safety, health, and environmental process is to be improved its performance must be carefully measured and monitored. Various safety, health, and environmental performance indicators, both lagging and leading, can be measured and charted to assess the functioning of the SHE process. Control charting is commonly used to graph traditional lagging indicators such as accident frequency rate and the number of accidents per time period in order to assess the true nature of accident trends. Application of the correct SPC charting method enables the SHE pillar to determine whether month-to-month or quarter-to-quarter changes in accidents are due to natural variation of the safety process, to random cause, or to real improvements in the safety process. Table 7.3 outlines which type of control chart should be used for tracking different types of attributes and variables, and provides the formulas for key SPC chart parameters. As indicated in the table, U-charts are attribute charts that are used for tracking normalized rates, such as accident and incident rates, whereas C-charts are used for measuring events or counts such as numbers of accidents or incidents. X-charts can be used for charting individually measured quantities or variables.[10]

To increase the understanding of SPC charting and to improve decision making, color-coded dashboards can be posted along with the charts summarizing the current trend in SHE performance. For example, the following color-coded system, similar to those recommended by Prevette and others,[11,12] could be used to quickly communicate to management and employees the status of the various safety, health, and environmental parameters that are being charted:

Dark Green: Performance is stable and at a superior level, exceeding target.
Blue: Performance is stable and at an acceptable level, meeting target.
Orange: Performance is stable but not acceptable, not meeting target.
Red: Trend due to special cause is adverse.
Light Green: Trend due to special cause is favorable.

Organizations that desire sustained success in the safety, health, and environmental area should also track leading indicators to enable them to understand SHE performance better and to facilitate proactive action to address adverse trends before they become a problem. In the safety area, leading indicators include routine weekly or monthly tracking of the number of unsafe conditions or SHE F-tags, at-risk behaviors, near misses, inspection and survey findings, audit points, training sessions conducted, and safety activities completed. In the occupational health area, leading indicators include exposure levels, results of medical surveillance monitoring, number of persons removed from exposure, findings of health assessments and audits, and completion of health promotion activities. In the environmental area, leading indicators include environmental near misses, results of environmental monitoring, and completion of environmental activities.

Table 7.3 Formulas for Different Types of SPC Control Charts and Their Uses

Chart Type	Use	Formula for Mean	Formula for Standard Deviation	Formula for Control Limits
C-Chart	Charting of events or counts For C-chart: n = number of months X_t = no. events reported per month Σ from t = 1 to n	Formula for mean when charting no. of events per period: $\bar{x} = \dfrac{\sum x_1 + x_2 + \cdots x_n}{n}$	Formula for std deviation when charting no. events per period: $\sigma = \bar{x}^{1/2}$ where $\bar{x} > 5$	$UCL = \bar{x} + 3\sigma$ $LCL = \bar{x} - 3\sigma$
U-Chart	Charting of nonconformities when sample size varies such as normalized rates For U-chart n = number of months x_t = no. cases reported/mo. h = work hours h_t = total work hours Σ from t = 1 to n	Formula for mean when charting OSHA recordable rate: $\bar{u} = \dfrac{\sum x_1 + x_2 + \cdots x_n}{\sum h_1 + h_2 + \cdots h_n/200{,}000}$	Formula for std deviation when charting OSHA recordable rate: $\sigma = \left(\dfrac{\bar{u}}{h_t/200{,}000}\right)^{1/2}$	$UCL = \bar{u} + 3\sigma$ $LCL = \bar{u} - 3\sigma$
X-Chart	Charting of variables/measured quantities For X-chart: n = no. of time periods x_t = value for period $d_2 = 2/(3.14)^{1/2}$ Σ from t = 2 to n	$\bar{x} = \dfrac{\sum x_1 + x_2 + \cdots x_n}{n}$	$\sigma = 1/d_2\left(\dfrac{\sum |x_t - x_{t-1}|}{n-1}\right)$	$UCL = x + 3\sigma$ $LCL = x - 3\sigma$

7.5 SHE Visual Mapping and Charting

A key principle of Lean is that visual methods enable employees to better understand and manage workplace hazards and risks. Indeed, there is a lot of truth in the cliché that proclaims a picture is worth a thousand words. Mapping techniques are useful and effective for studying safety, health, and environmental issues because they enable teams to look at and analyze their workplace in a new way. With this fresh perspective, employees are able to gain new insights and are better able to develop collaborative solutions that really work. With this in mind, the SHE pillar team, work area teams, and *kaizen* teams should be encouraged to apply visual mapping methods to improve the assessment and control of a wide variety of workplace hazards. Examples of visual mapping and charting methods useful in the SHE area include dot distribution maps, body maps, fishbone charts, discharge maps, process flow diagrams, and contour maps.

7.5.1 Dot Distribution Mapping

The placement of dots on area or equipment drawings indicates the location, distribution, and concentration of workplace hazards or environmental impacts. These dot or measles maps can be created from completed equipment F-tags, the findings of area inspections, and the results of safety, health, and environmental studies. Figure 3.7 in Chapter 3 provides an example of a measles map generated from F-tag data. This type of visual mapping can assist employee teams in identifying the source of safety, health, and environmental abnormalities or hazards.

7.5.2 Body Mapping

This visual method involves marking the location of all reported work-related injuries, illnesses, symptoms, or stresses on a diagram of the human body. As shown in Figure 7.6, the body map provides a clear picture of the types of accidents experienced and the body parts affected by workplace accidents and stresses. The body map can be used to convey to employees important information about the types of accidents, the body parts most frequently involved in workplace accidents, and workplace risks. Body mapping can also be useful for studying workplace health complaints. Employees are asked to mark the body diagram with the location and type of health complaint or symptom they are experiencing. Unique colors are

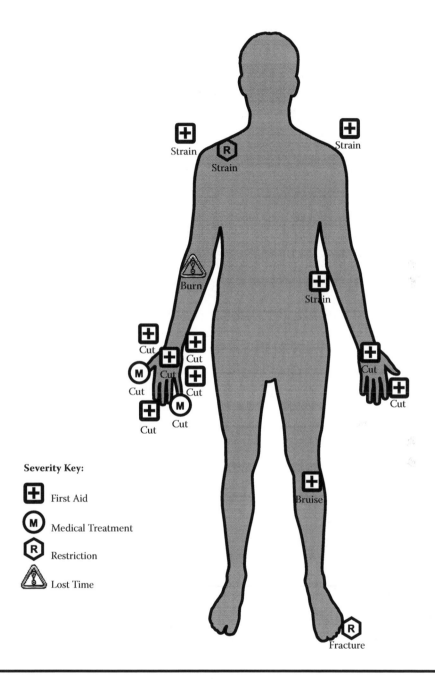

Figure 7.6 Example of a body map showing injury distribution.

used to mark different types of injuries, illnesses, and symptoms. When completed, the body map provides a picture of the extent and pattern of workplace complaints. This type of body mapping is often useful for studying workplace complaints related to ergonomics or exposures to chemicals and physical agents.

7.5.3 Fishbone Mapping

Fishbone maps or Ishikawa diagrams are cause-and-effect diagrams that show the causal relationship between various factors and an outcome. As shown in Figure 7.7, the causal factors are commonly divided into several broad categories known as the 5Ms: man or people, materials, methods, machines or equipment, and milieu or environment. Fishbone diagrams can be used to structure brainstorming session analyzing the causes of safety, health, or environmental problems, and to display the results of the analysis.

In behavioral or behavior-based safety, fishbone diagrams can be used to analyze the factors that influence and cause employee behavior. Human behavior is influenced by both antecedents, things that come before the behavior, and consequences, the real and perceived outcomes of the behavior. Therefore, in analyzing employee safety-related behavior a two-way fishbone that evaluates both the antecedents and consequences of employee actions can be constructed. This two-sided fishbone approach for analyzing workplace behaviors is commonly known as A–B–C analysis or antecedent–behavior–consequence analysis.

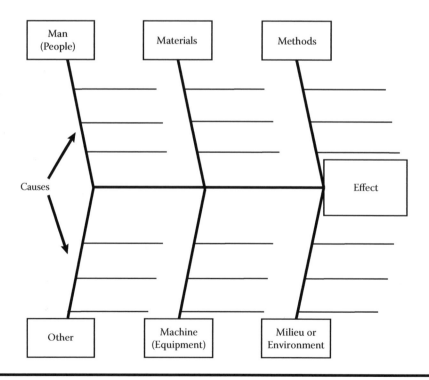

Figure 7.7 Standard fishbone cause-and-effect diagram.

As shown in Figure 7.8, this methodology can be used to identify the reasons why employees may be engaging in an unsafe act or at-risk behavior such as failure to wear personal protective equipment (PPE) when unloading acid. At first glance it may seem foolish not to wear PPE when handling a hazardous material, and failure to do so may appear to be a simple act of employee misconduct. However, humans are rational beings and as such they take certain actions, even risky ones, because it makes sense to them at the time. By using A–B–C analysis, upon closer evaluation it becomes clear that there are many reasons why an employee may take a safety risk. Both antecedents and consequences influence employee risk-taking. In the example shown, several antecedents influenced and contributed to the employee's decision not to wear PPE. The antecedents included the following factors: the employee was not trained, and because the worker was new to the company he was unfamiliar with the task and did not fully understand the job hazards. In addition, there were no written job procedures, and inasmuch as it was a very hot day the employee was inclined not to wear the heavy and uncomfortable PPE. This evaluation of the antecedents influencing the employee's unsafe behavior enables the

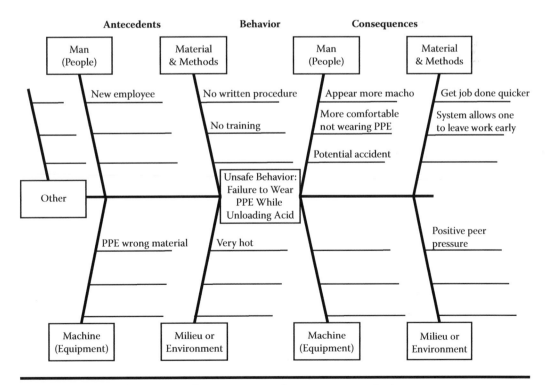

Figure 7.8   Use of fishbone diagram to conduct A–B–C analysis.

organization to identify actions that must be taken to promote the desired safe work practices.

The right side of the A–B–C analysis is a listing of the real and perceived consequences of the risky behavior from the employee's perspective derived from a brain-storming session. As part of A–B–C analysis each of these consequences should be evaluated from the employee standpoint by considering the likelihood of occurrence, how soon it will occur, and if the consequence is positive or negative. Consequences that are certain, soon, and positive tend to be the most powerful and have the most influence on employee behavior. Of course, one possible consequence of the failure to wear PPE when handling a hazardous material is an injury. But from the employee's perspective this negative consequence is unlikely because it has never happened to him before, and besides the worker often believes that he will avoid injury by being careful when doing the job. From the worker's perspective, failure to wear the bulky PPE on a hot day does have some immediate, certain, and positive consequences that are quite powerful. It is these "soon–certain–and positive" consequences that are driving the risk-taking. From the employee point of view, not wearing the PPE results in more comfort, the ability to complete the job quicker, and the opportunity to leave work early. After evaluating and understanding the significance of each consequence from the employee's perspective, the organization is then able to encourage safe work practices by restructuring its management system to enhance the payoffs for safe behavior. The safety learning derived from A–B–C analysis can be posted sitewide to ensure that all employees benefit from the insights gained.

7.5.4 Discharge Mapping

Development of site emission and effluent maps depicting the location, source, type, and magnitude of facility discharges is useful for environmental education and problem solving. These color-coded maps and in-field markings that identify each environmental discharge point via the use of unique colors and numbers increase the understanding of a site's potential environmental impacts and aid in identifying root causes and solving problems when issues develop. The site's SHE pillar team and SHE professionals can work with area teams to develop environmental maps for each area and process. These maps can be posted in the work area to increase understanding of the organization's environmental footprint and to aid in the identification of opportunities for environmental improvement.

7.5.5 Process Flow Mapping

Process flow maps are block diagrams showing the steps of an activity, process, procedure, or decision and their interrelation. In the SHE area, process flow diagrams have a wide variety of uses including

1. To understand the sequence of steps in a supply chain process and the activities associated with the potential generation of and exposure to safety, health, and environmental hazards
2. To analyze the flow, adequacy, and effectiveness of the steps in a SHE procedure
3. To communicate to others the steps and SHE issues in a process or procedure in order to increase understanding and to ensure safe and efficient operation

Figure 7.9 provides an example of a chemical process flow diagram with pictographs providing a visual indication of what types of hazards exist at each step in the process. Diagrams of this type are useful for training new

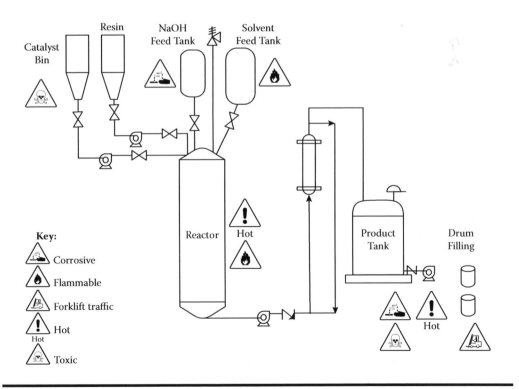

Figure 7.9   Simple chemical process flow diagram with safety symbols.

employees on the hazards of the process, and for providing a periodic visual safety reminder to current workers.

In many business processes there are often multiple steps carried out by different people and different functions resulting in handoffs of responsibilities and duties that are often a source for mishaps and delays if not carefully managed. To ensure that the business process is fully understood and effectively implemented, a special type of process flow map, known as a swim lane diagram, can be used to clarify roles and responsibilities. As shown in Figure 7.10, a swim lane diagram divides the process flow into separate parallel lanes depicting what person or function is responsible for specific process steps or subprocesses. Developing a swim lane diagram for a SHE process, such as the SHE F-tag procedure shown in Figure 7.10, is helpful in identifying weaknesses and unnecessary delays in the process. Because the example swim lane diagram shows that the main delays in repairing substandard workplace conditions occur in the Maintenance Department, this department's procedures should become the focus of study and *kaizen* activity. Consistent with basic Lean principles, data on process delays and losses that are derived from flowcharts are used to drive continual improvement.

7.5.6 Contour Mapping

Contour lines are commonly used in topographical maps to divide a given geographic area into zones of equal elevation. Contour mapping can also be used in the SHE field to create an informative visual map of the noise, heat, radiation, or chemical exposure levels in a particular workplace or external environment. As shown in Figure 7.11, a noise map uses contour lines to connect points of equal average noise level, thereby creating a picture of the various noise zones in a workplace. Noise contour maps are useful for assessing noise exposure levels, determining the sources of high noise levels, and for tracking the success of noise control efforts over time. With the assistance of SHE professionals, the site SHE pillar team, area teams, and noise *kaizen* teams can create and post noise maps of the work area. The maps enable one to quickly identify the high noise areas on a production line, and to pinpoint the specific equipment sources of elevated noise levels.

In the example provided in Figure 7.11, the noise contours clearly indicate that the case packer is a key high-noise source. By taking additional sound level readings and drawing additional noise contours at higher noise levels, it

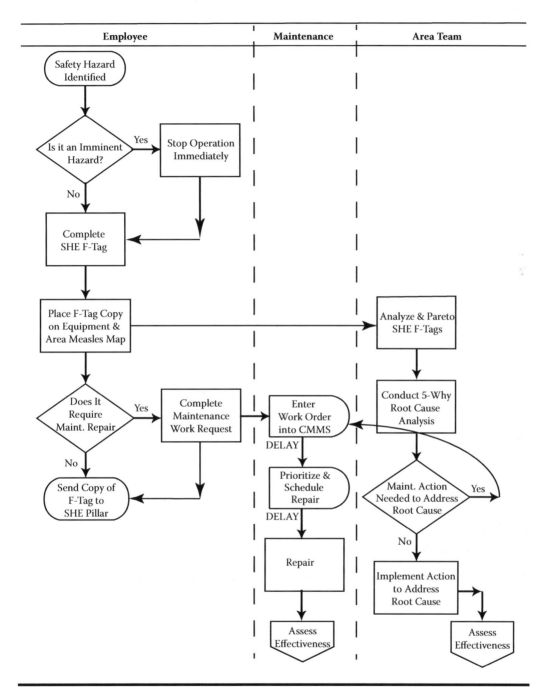

Figure 7.10   Swim lane flowchart of a SHE F-tag procedure.

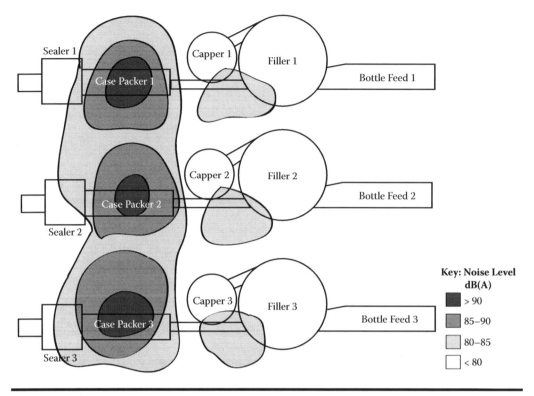

Figure 7.11 Example of a noise contour map.

is possible to identify the specific sources of the noise. Careful analysis will enable the team to determine whether the elevated noise levels of the case packer are the result of uncontrolled air discharges, impact noises due to the metal-to-metal contact of machine parts, or some other cause. By comparing the noise maps from the current time period to those of previous time periods, a clear picture of the progress made in the site's noise control program is obtained.

7.6 SHE Pillar Activity Board

Activity boards are used in Lean/TPM implementation to communicate the status of pillar activities and the progress of *kaizen* or improvement projects. In addition to their fundamental communication role, activity boards perform an important training function and also serve to structure and order the thinking and problem-solving approach of employee teams. The pillar activity board is not meant to be an arts and crafts project, but rather a tool for visually communicating, reinforcing, and embedding the Lean/TPM continuous improvement process throughout the organization. The boards are posted in the workplace for all to see, and periodically the pillar teams and *kaizen* teams are asked to make formal presentations on the content of their activity boards.

Like all other pillars, the organization's SHE pillar team should construct an activity board or boards that clearly explains the process the site is using to control SHE risks and to eliminate SHE accidents, incidents, and losses. The SHE pillar board serves to educate all employees on the Lean tools and methodologies the site is using on its journey toward triple zero: zero accidents, zero incidents, and zero losses. Figure 7.12 provides an example of an effective way to structure a safety and health activity board. Due to the wide scope of SHE activities, many organizations choose to construct both a safety and health activity board and a separate environmental sustainability activity board.

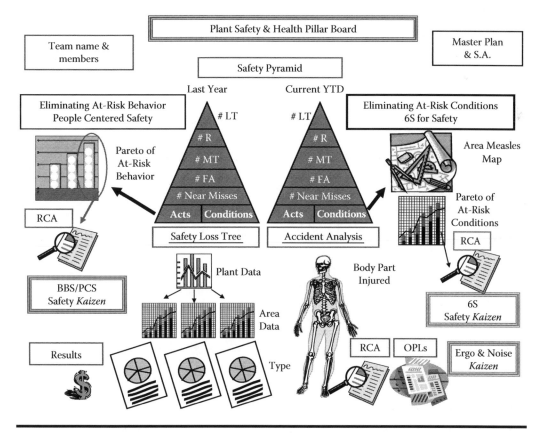

Figure 7.12 Example of a safety and health activity board.

The top of the safety and health activity board lists the team's moniker and the names of its members, and provides a summary of its pillar plan and a copy of its latest self-assessment radar chart. Conceptually the board is divided into three sections. Because all safety activities should be driven by the site loss tree, the central portion of the board displays important data on the site's safety and health losses. Two segmented triangles based on Heinrick's safety pyramid are used to compare the site's current year and previous year accident and incident data. Significant emphasis should be placed on the bottom of these safety pyramids, which display information on the site's leading safety and health indicators such as at-risk behavior and at-risk conditions. The central section of the board further analyzes the site's loss tree data with pie charts displaying the locations and types of accidents, and a body chart visually depicting the severity of accidents and the body parts injured. The results of accident root cause analyses (RCAs) are posted along with safety one-point lessons (OPLs).

To be excellent in safety the organization must be effective at identifying and removing at-risk conditions and at-risk behavior from the workplace. A key role of the site SHE pillar team is overseeing and facilitating this risk elimination process, thus, the rest of the activity board is devoted to telling the story about the site's safety improvement efforts. The right side of the board explains the site's 6S process for eliminating unsafe conditions. A measles map graphically communicates the location of at-risk conditions, and a Pareto chart shows the site's most common at-risk conditions. Below this the activity board describes the *kaizens* the site has implemented to address the unsafe conditions.

The left side of the activity board explains the behavior-based safety (BBS) process the organization has used to systematically identify and eliminate at-risk behavior. Site safety observation data are charted using Pareto analysis that clearly indicates the most common at-risk behaviors. Below the analysis of the organization's unsafe behaviors is information describing the *kaizens* that have been undertaken to eliminate specific at-risk behavior and to promote safe work practices. At the bottom of the board the results of the organization's safety and health improvement initiatives are summarized outlining the reduction in accidents, incidents, and losses, highlighting both the decline in the number of cases and the decrease in costs. In this way, the safety and health activity board serves as a training tool and as a public proclamation of the progress the organization is making on its journey toward safety excellence by using Lean methodologies.

Endnotes

1. Rule Book of the New York and Erie Railroad (1854). "Great Aviation Quotes." Retrieved on Sept 29, 2010 from: http://www.skygod.com/quotes/safety

2. Manuele, Fred A. (August, 2007). Lean concepts: Opportunities for safety professionals. *Professional Safety.* 52, 8: 28–34.

3. Hallowell, Matthew, Anthony Veltri, and Stephen Johnson (November, 2009). Safety & Lean; One manufacturer's lessons learned and best practices. *Professional Safety.* 54, 11: 22.

4. ISO (2004). ISO 14001:2004 Environmental management systems—Requirements with guidance for use. International Organization for Standardization. Geneva, Switzerland.

5. OHSAS (2007). OHSAS 18001-2007: Occupational health and safety management. Retrieved from: http://shop.bsigroup.com/en/ProductDetail/?pid=000000 000030148086

6. Gawande, Atul (2009). *Checklist Manifesto: How to Get Things Done Right.* Metropolitan, New York.

7. Hutchens, S. (2010). "Using ISO 9001 or ISO 14001 to gain a competitive advantage." Intertek white paper. September 2010. Retrieved from: http://www.intertek.com/WorkArea/DownloadAsset.aspx?id=4431

8. Owen, Mal (1989). *SPC and Continuous Improvement.* IFS, Kempston, UK, pp. 99–104.

9. Owen, op. cit., pp. 115–120.

10. DOE (1992). DOE performance indicators guidance document. DOE standard 1048-92. U.S. Department of Energy, Washington, DC.

11. Prevette, S. (2006). Charting safety performance: Combining statistical tools provides quality data. *Professional Safety.* 51, 5 (May): 34–41.

12. Prevette, S., W. Previty, A. Umek, and C. Hayes, (2009). Implementing performance trending at two Department of Energy sites. *WM Conference*, March, 2009.

Chapter 8

Lean and Green: Applying Lean to the Environment

The world will not evolve past its current state of crisis by using the same thinking that created the situation.

Albert Einstein

8.1 Lean and Environmental Sustainability

Lean production and environmental sustainability share the same overall goals and objectives: to eliminate waste, to maximize the use of resources, and to achieve long-term efficient operations. In most manufacturing processes air pollution, water effluent, and waste are composed of valuable raw materials and finished goods that don't make it into the product, but instead are discharged into the air, flushed down the drain, or buried in the ground. Environmental sustainability or "Green" efforts make business sense because they focus on the efficient use of resources, which has a positive and salutary effect on an organization's finances. Contrary to conventional wisdom, there is indeed compelling evidence that those companies can be both Lean and Green.

Business goals and environmental objectives are not automatically antagonistic and mutually exclusive but, rather, can be complementary and even synergistic when properly aligned and implemented. In the book, *Lean and Green*, Pamela Gordon[1] argues that business does not have to choose between environmental sustainability and profit. The author provides convincing evidence gathered from business case studies to dispel the myth that Green practices are contrary to business success. Gordon cites companies such as Texas Instruments, IBM, and Intel that have significantly reduced costs by reducing wastes. Other companies such as Horizon Organic Dairy, the largest U.S. supplier of organic milk products, have increased revenues by developing "green" products and services that address consumer needs. In the end, successful sustainability initiatives must make both environmental sense and business sense or they will not endure. Thus, while protecting the environment and preserving natural resources, they should also deliver long-term profitable growth.[2]

Businesses can achieve both environmental and productivity improvements by making their processes more efficient (eco-efficiency), by developing more environmentally friendly products (eco-innovation), and by operating in a more sustainable fashion. Lean and Green progress can be achieved via incremental improvements in existing ways of working or by making fundamental changes in one's business model and systems. In fact, many environmental advocates including Paul Hawken, author of *The Ecology of Commerce* and *Natural Capitalism*, argue that current business practices are devastating the planet, and therefore significant and lasting sustainability progress can only be made via a fundamental redesign of business models and commerce.[3,4] The solution is to establish Lean processes and a "restorative economy" that are designed to mimic nature by eliminating, reusing, or recycling all waste. Business should abandon the decidedly wasteful and un-Lean "cradle to grave"[5] approach where a business's wastes are dumped into the environment. Instead, nature's "cradle to cradle" system, where the concept of waste does not exist, should become the basis of industrial design. In nature, one process's wastes are reborn by becoming raw material for another process. Thus, in a "cradle to cradle" economy old products are not buried in the grave, but instead are repeatedly given new life by being restored by the manufacturer, or are reincarnated as an alternative beneficial resource or product.

The Lean and Green approach to sustainability demands new ways of thinking and acting. The old ways of designing and producing products will no longer suffice. The Lean early management pillar and the SHE pillar can work together to brainstorm and evaluate options for sustainable smart products and processes. These smart designs should be based upon nature and use natural resources, or natural capital, wisely, and beneficially reuse all wastes. Successful Lean and Green sustainable solutions will require new product, production, and distribution paradigms such as

- Paints without solvents
- Vehicles without gasoline
- Cleaners without surfactants
- Agriculture without pesticides
- Food produced with less water
- Refrigeration without refrigerants
- Ice cream without refrigeration
- Nuclear power without waste
- Power without transmission loss
- Power without carbon and pollution
- Energy without fossil fuels
- Zero waste manufacturing
- Disposable products replaced by reusable products

A few companies such as Interface Global, the world's largest producer of modular carpet, have established a business model based on environmental sustainability and Lean principles. At the urging of its founder and chairman, Ray Anderson, Interface has adopted a new business paradigm that aims to design and operate industrial and commercial processes in the way that nature operates.[6] Rather than disposing of waste, reuse it. Rather than consuming limited energy resources, use renewable energy. Rather than depleting natural resources, use them in a sustainable fashion. With nature's renewable systems in mind, Interface abandoned the traditional carpeting business model where customers dispose of old carpeting and purchase new carpet made of virgin materials. Instead, Interface has employed a business model that promotes the return and recycling of carpeting. Modular carpeting products are manufactured in sustainably focused facilities that utilize significant quantities of recycled and renewable raw materials. The company's recycling program allows customers to be active participants in the company's environmental efforts by returning their old carpet tiles to be recycled into new products. Carpeting that was once destined to clog landfills is now used as feedstock for new carpet. Interface's corporate vision statement summarizes the organization's fundamental commitment to operating its business in an environmentally sustainable manner:

> To be the first company that, by its deeds, shows the entire indus-
> trial world what sustainability is in all its dimensions: People, pro-
> cess, product, place and profits—by 2020—and in doing so we will
> become restorative through the power of influence.[7]

Rather than making minor environmental improvements at the margins of its business, Interface has focused on big Lean and Green improvements by rethinking and redesigning its way of working and doing business. This approach has delivered both business success and significant environmental improvement for Interface:[8]

- From 1995 to 2008 waste reduction activities have saved $405 million, reduced landfill disposal by 75%, and diverted 175 million pounds of material from landfills.
- The company has reduced energy use at carpet manufacturing facilities (per unit of product) by 44% since 1996.
- 28% of the company's total energy use is derived from renewable sources.
- Interface's net greenhouse gas emissions are down 71% from a 1996 baseline: 34% as a result of energy efficiency and direct purchases of renewable energy, and 37% offset by company-owned GHG projects.

Reflecting on Interface's sustainability journey Ray Anderson proclaims that "sustainability has given my company a competitive edge ... it has proven to be the most powerful marketplace differentiator ... our costs are down, our profits are up, and our products are the best they've ever been."[9] Interface's experience has demonstrated that a Lean and Green approach can deliver profitable and sustainable growth.

John Elkington's[10] concept of the triple bottom line argues that modern organizations should not only focus on their traditional financial bot-
tom line, but should also consider the social and environmental impacts of their activities. As shown in Figure 8.1, sustainability can be seen as the intersection of an organization's efforts to manage the social (people), environmental (planet), and economic (profit) impacts of its activities.[11] In other words, organizations are responsible and accountable for the 3Ps: people, planet, and profit. Integrating a company's Lean and sustainability

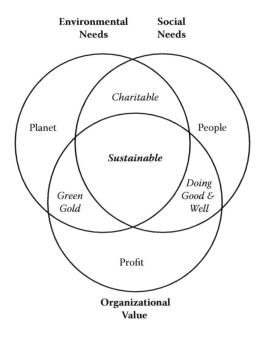

Figure 8.1   The sustainability triple bottom line: people, planet, and profit.

or Green processes provides an effective mechanism for managing the 3Ps. As shown in Figure 8.2, Lean provides a systematic methodology for managing organizational demands, such as the 3Ps, while delivering a triple-value deal of customer value, internal organizational value, and increased responsiveness.

According to James Womack, "Lean thinking must be 'Green' because it reduces the amount of energy and waste by-products required to produce a given product. ... Lean's role is to be Green's critical enabler."[12,13] Whereas Lean focuses on understanding and meeting customer needs and maximizing value, Green concentrates on satisfying societal and environmental needs while maximizing value. In today's increasingly fragile world with its growing number of environmentally conscious consumers; society's, the customer's, and the environment's needs often intersect. The company that successfully integrates its Lean and Green initiatives is poised to reap long-term rewards in the marketplace.

The sustainability era is upon us. Even the *Harvard Business Review* has proclaimed that sustainability is an emerging megatrend that businesses ignore at their own peril.[14] Furthermore, a recent United Nations and Accenture report entitled "A New Era of Sustainability" reports that 93% of CEOs believe that sustainability issues will be crucial to the future success of

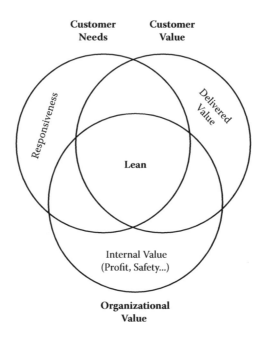

Figure 8.2   The Lean triple value deal: delivered value, internal value, and responsiveness.

their companies.[15] Thus, the issue is not whether businesses should adopt a sustainable approach, but how they should do so. A strategic approach that integrates the organization's sustainability or Green process with the organization's Lean management system to deliver customer value is a recipe for continual and enduring business success. The Lean and Green organization is able to "do well by doing good," in a sound and sustainable fashion.

8.2 The Lean and Green Sustainability Roadmap

The first step in the Lean and Green journey is to develop and adopt a clear sustainability roadmap. Like any map, the Lean sustainability roadmap should provide the organization with a clear path to its final destination including key intermediate stops and mileposts along the way. Figure 8.3 provides a 10-step model that organizations can use to chart their paths along the road to sustainable and profitable operations. Although each organization must chart its own unique path to building a truly sustainable and successful business, this model provides an outline of the key steps that one should consider to ensure a successful sustainability journey.

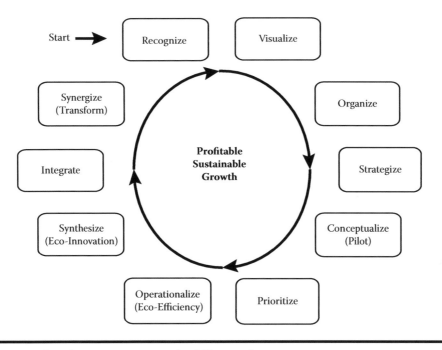

Figure 8.3 **The virtuous Lean sustainability cycle.**

8.2.1 Phase 1: Recognize

The starting point for a business's Lean and Green quest is recognizing the strategic need and opportunity presented by sustainability. It is essential that the organization's leadership see and believe that there is a business imperative for pursuing an often long and challenging Lean and Green sustainability journey. Leadership must clearly understand the need and the business case for sustainability. Senior management should publicly declare the need for change, and the strategic necessity of rethinking and re-engineering the organization's ways of working in a leaner and more sustainable fashion. Leadership must be committed to re-examining the company's business model, management systems, and operating methods with both a Lean and a sustainability lens. Ultimately, the success of the Lean and Green initiative depends upon leadership's commitment and engagement because they define the organizational strategy, establish priorities, assign resources, and bestow rewards. As the old management maxim says, "What gets measured gets done; what gets measured and rewarded gets done well." Leadership's key role is to create an organizational environment that encourages, fosters, and rewards Lean and sustainable business practices.

8.2.2 Phase 2: Visualize

The second part of the Lean sustainability journey involves visualizing the current state, the challenges ahead, and the path forward. In this phase the organization should establish Lean and Green measurement systems that enable the organization to gather baseline and ongoing data on sustainability and efficiency. A clear picture of the current state of the organization's business practices and sustainability performance is obtained. In this stage leadership should also be actively engaged in visualizing and formulating the organization's Lean sustainability vision. The organization's Lean and Green vision should be more than a lofty statement that is proclaimed and quickly forgotten. Rather, the vision must be an inspiring call to action that provides a clear and compelling vision of the desired future state. By comparing the organization's current state with its Lean and Green vision, a clearer understanding emerges of the performance gaps that need to be closed to make the Lean sustainability vision a reality.

8.2.3 Phase 3: Organize

Once the organization's Lean sustainability vision is established, leadership must organize the enterprise to achieve this vision. As in nature, form follows function. If the organization desires to function in an efficient and environmentally sound fashion, its "form" or organization must be designed with Lean and sustainability in mind. The organization's resources, personnel, and processes must be rethought and re-engineered to support the Lean and Green vision by building capability and competency in this area. Ideally, the Lean sustainability process is owned by senior management, and to ensure a clear and consistent focus it is led by an officer, director, or senior manager knowledgeable in Lean and sustainability. Expert internal and external input must be sought to ensure that the organization's efficiency and environmental efforts are based upon sound principles and do not degrade into a superficial public relations and lip service exercise characterized by "greenwashing" and business as usual. Consideration should be given to how best to structure and resource the organization in order to maximize its eco-innovation and eco-efficiency efforts, and to capitalize on the business opportunities presented by sustainability.

8.2.4 Phase 4: Strategize

A detailed strategy on how to deliver the organization's Lean and Green vision should be formulated. This Lean and Green strategy should not be a public relations effort focused on promoting the company image but, instead, a broad business plan that concentrates on real stratagems and approaches that can deliver sustainable profitable growth. When attempting to craft a successful Lean sustainability strategy, organizations are well served to concentrate on sustainability issues and initiatives

1. Relevant to the company's core businesses
2. That can be leveraged to build brands and grow the business
3. Where the business can exert market influence and bring about real lasting change
4. Associated with reducing the organization's significant sustainability impacts
5. Where the organization has special interest and expertise

8.2.5 Phase 5: Conceptualize

Lean and sustainability are not only about vision and strategy; they are also about execution. Ultimately, long-term Lean and sustainability success is the product of strong strategy and effective execution. To ensure that the Lean and Green efforts are as effective and efficient as possible, it is often helpful to test and use trial initiatives before launching them enterprisewide. In the conceptualization phase, pilot projects are conducted to test out and prove the organization's various approaches to achieving sustainable and profitable operations. Pilot projects addressing low-hanging fruit in the areas of recycling, reuse, and resource conservation enable organizations to demonstrate early success and develop confidence in their new Lean and Green ideas.[16] During this phase the company derives important learning from these isolated Lean and Green initiatives regarding which approaches hold the most promise for organizational success and transformation.

8.2.6 Phase 6: Prioritize

After testing out and proving various approaches, the organization is prepared to develop a more exhaustive operational plan that details specific priorities, responsibilities, and timelines for implementation across the enterprise. No organization has unlimited resources, and so it is essential to focus efforts where they will be successful and have the most impact. During these early stages, the time spent planning, piloting, and prioritizing the Lean and Green initiatives is well spent, because it increases the chances of success and ensures that the effort is focused on the most important sustainability and efficiency issues.

8.2.7 Phase 7: Operationalize

As the Lean and Green process matures, the organization begins to incorporate Lean and sustainability into its core business operations and its ways of working. As suggested by David Lubin and Dan Esty, initial efforts generally focus on eco-efficiency by optimizing existing processes and "doing old things in new ways."[17] As a first order of business it is important to get one's own house in order by increasing process and equipment efficiency, reducing waste, and eliminating unsound environmental practices in existing operations.

8.2.8 Phase 8: Synthesize

As the organization increasingly incorporates Lean and sustainable thinking and practices into existing operations it then turns its attention to applying them to new operations. Lean and Green efforts concentrate on widespread eco-innovation by redesigning processes and products. The synthesis of Lean and Green thinking results in new products and processes with improved environmental and efficiency performance.

8.2.9 Phase 9: Integrate

At this stage the Lean and Green process is fully integrated across the organization's operations including research and development, supply chain, and marketing. The practice where isolated functional silos engaged in separate, uncoordinated, and often unrelated sustainability projects has given way to a coordinated and integrated Lean sustainability process that aims to maximize company, customer, consumer, and environmental benefits. At this point Lean and sustainability are truly part of the organization's DNA, fully integrated into its way of doing business.

8.2.10 Phase 10: Synergize

During this phase, the organization leverages its Lean sustainability DNA to gain competitive advantage and to add value across and beyond its extended supply chain. The Lean sustainability process becomes the source of innovation and differentiation, giving birth to new business models, new products, and new services. The organization exploits its Lean sustainability process to build brands, expand markets, and grow profits. The organization extends its Lean sustainability process to include suppliers, customers, and consumers, fundamentally changing the way that it does business. The business's sustainability efforts expand beyond reducing environmental impacts to actual restoration of the environment. The company has established a virtuous cycle where it is able to build a growing, profitable, and sustainable business. Because Lean and Green thinking is firmly embedded in the organization it is able to recognize, respond to, and capitalize on new and emerging sustainability challenges.

8.3 Applying Value Stream Mapping to the Environment

Value stream mapping is a methodology for visually charting the sequence and timing of activities, and the flow of materials and information required to produce a product or service. The value stream map is commonly utilized in Lean implementations to provide a visual picture of value generation and waste production in a process in order to identify opportunities for improvement. However, because the value stream mapping often focuses exclusively on the internal workings of a process, it effectively identifies production-related wastes, but commonly fails to identify environmental wastes. By expanding the focus of the value stream mapping exercise, it can be extended to the environmental sustainability area to identify opportunities for reducing environmental discharges and waste, and for conserving resources.

Before beginning environmental value stream mapping it is helpful to have visibility and an understanding of the entire value chain. As shown in Figure 8.4, the value chain of a typical product or service extends beyond one's internal production processes to include the sourcing activities of upstream suppliers and transporters, the operations of downstream distributors and customers, and ultimately the actions of the consumers when they use and dispose of the product or service. During each step of this extended value chain wastes are produced and environmental impacts occur. In fact, life-cycle analysis has

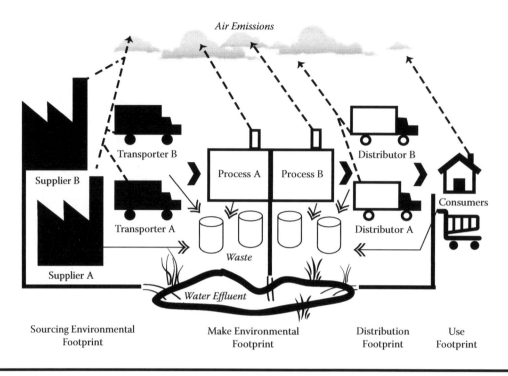

Figure 8.4   Environmental footprint of the extended supply/value chain.

indicated that for many consumer products the largest environmental impacts do not occur at the make stage, but rather during the sourcing of raw materials and the final use and disposal of the product by the consumer.

Apple, for example, reports that 53% of the greenhouse gas emissions from its products occur during consumer use.[18] Unilever, a large multinational manufacturer of food, personal care, and home care products, has reported that 69% of the carbon footprint of its products is generated through consumer use and disposal, 26% is generated from raw material sourcing, and only 5% from manufacturing and distribution.[19] To effectively control the total environmental footprint associated with their product, organizations must understand the product's complete life cycle and its entire environmental value chain, accounting for environmental impacts arising from resource consumption, production, distribution, use, and disposal. Establishing systems that collect data on key environmental metrics at each stage of the value chain will enable an organization to quantify the environmental impacts at each step and will facilitate good decision making on reducing a product's overall environmental footprint.

Environmental value stream mapping can and should be used at each stage of a product's value chain or life cycle. Figure 8.5 provides an

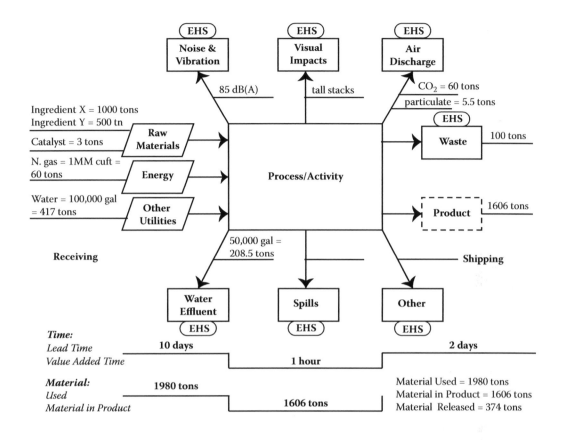

Figure 8.5 Environmental aspect and value chain flowchart.

example of how environmental value stream mapping can be applied to one process or activity in a value chain. The first step in constructing the environmental value stream is for a multidisciplinary team of persons knowledgeable in the process under study and the associated environmental issues to identify and quantify all inputs to the process. Typical process inputs include raw materials, catalysts, energy, and other utilities such as air and water. Next the intended process outputs, the finished products, should be identified and quantified. Few, if any, processes are 100% efficient in converting raw materials to finished product. Unfortunately, some raw materials or process inputs are not transformed into product, but rather are wasted and released to the environment. Thus, the final and most important step in constructing the environmental value stream map is to identify and quantify all the processes' environmental, health, and safety outputs including waste, water effluent, routine air discharges, spills, and noise and visual impacts.[20]

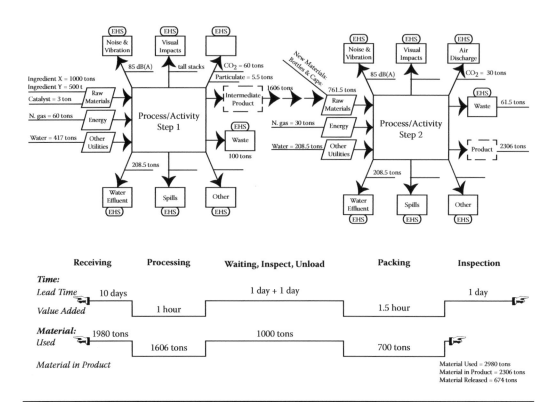

Figure 8.6 Environmental value stream map of a two-step process.

As shown in Figure 8.6, environmental value stream mapping can be applied to multistep processes by simply constructing a value stream fishbone diagram for each step in the process. Generally, the output of one process becomes the input of the next process. A well-constructed current-state environmental value stream map clearly identifies all environmental, health, and safety aspects and impacts of an existing process by using a unique symbol such as EHS or SHE to represent these impacts. The environmental data for the process is recorded on the map to enable quantification and assessment of these impacts. Data to enable quantification of environmental wastes, releases, discharges, and impacts can be obtained from engineering estimates, environmental monitoring, measurements on the shop floor or *gemba*, or by conducting in-process studies.

Creation of the environmental value stream map provides a rough material balance for the process, enabling one to understand how efficient the process is in converting raw materials to the desired finished product, and where there are opportunities to reduce waste and environmental impacts. The material chart at the bottom of the value stream map shows on the top line the amount of material that is used in the process versus the amount of material on the bottom line that is transformed into product. The amount of finished product represents value to the customer, whereas unconverted material represents waste, potential impacts to the environment, and improvement opportunities.

8.4 Applying Visual Controls to the Environment

Visual controls used in the production process, such as *kanban* systems that provide a clear signal to an upstream process regarding when to produce a desired component, serve to match product production with product demand. Thus, overproduction and its wasteful use of resources is prevented. The production leveling brought about by Lean methodologies such as *kanban* systems and just-in-time (JIT) manufacturing result in the reduction of several types of waste including

1. *Product waste:* The manufacture of excess product that is unneeded and unwanted. This overproduction leads to excess product inventory that ties up organizational capital, and will likely entail additional costs when it must be scrapped.
2. *Wasted raw materials:* The unnecessary consumption of valuable raw materials and human resources that could be put to better use.
3. *Energy loss:* The use of energy and other utilities to create nonvalued product.
4. *Environmental wastes:* The generation of waste, air emissions, and water discharges as a result of the overproduction.

Thus, in addition to their production benefits, traditional Lean visual control methodologies can have a positive impact on the environment and sustainability.

Visual controls and methods can be applied to the environmental area in order to reduce errors that lead to environmental incidents, and to increase awareness of and commitment to outstanding environmental performance. Visual controls should be applied to key tasks that have a potential environmental impact to ensure optimal and consistent performance of the task. Examples of visual controls and methods that can be utilized to enhance communication, motivation, learning, and ultimately action in the environmental area include

1. Color-coding, numbering, labeling, and mapping of all plant effluent lines and discharge points in order to enable the identification of the

source of all plant water pollution, and to facilitate problem solving related to effluent reduction and elimination

2. Color-coding, numbering, labeling, and mapping of all air emission points to enable the quick identification of the source of all air discharges and to aid pollution prevention and reduction efforts

3. Establishing unique color-coding for the containers of different types of facility waste in order to facilitate waste recycling and to prevent cross-contamination of waste

4. The labeling and color-coding of site spill supplies to ensure the quick and accurate response to spills, and thereby mitigating potential environmental impacts

5. The posting of large and visual one-point lessons (OPLs) for key environmental procedures to enhance learning and to reduce operator errors that can lead to environmental incidents

6. Displaying activity boards of environmental *kaizen* teams to communicate the progress toward implementing environmental improvements and related learning

7. Posting area and site environmental and sustainability metrics or key performance indicators and the progress toward meeting associated targets

8. Creating and posting control charts for key environmental variables such as critical process parameters that can affect the environment, effluent discharge data, and stack monitoring measurements

9. Applying 5S for SHE or 6S visual methods for environmental equipment and supplies to ensure that there is a clearly designated place for everything and everything is in its place

10. Employing *poka-yoke* mistake-proofing approaches to key environmental activities to ensure correct and consistent operator performance

8.5 Lean and Energy

There is a natural connection and synergy between Lean production and energy conservation programs: both disciplines are dedicated to eliminating waste and increasing process efficiency. By integrating Lean and energy-reduction efforts organizations can reap many significant benefits including

1. Reduced energy use and increased energy efficiency
2. Decreased release of greenhouse gases resulting in a smaller carbon footprint

3. Reduced discharge of conventional air pollutants including sulfur oxides (SOx), nitric oxides (NOx), volatile organic compounds (VOCs), and particulates
4. Lower emissions of toxic air pollutants
5. Cost savings and increased profitability

In the future, as fossil fuel supplies become scarce and demand for energy increases in the developing world, energy costs are expected to rise. Higher energy costs will increase the cost of doing business, and will place those companies that do not have a viable energy strategy at a distinct competitive disadvantage. Organizations that integrate their Lean and Green processes and develop a comprehensive energy strategy will reduce their costs, increase their competitiveness, and be more prepared to leverage opportunities in a carbon-constrained economy. With the goal of ensuring a secure and successful future for the organization, a company energy strategy should consider the following:[21]

1. The risks associated with the company's current and future energy supplies
2. The financial and business impacts of future energy price scenarios
3. The impacts of climate change regulations on the business
4. The effect of changing public opinion, consumer perceptions, and shareholder views of energy use on the business
5. The vulnerability of the business to fluctuating energy supplies, energy price volatility, the impacts of climate change, and changing societal and stakeholder concerns
6. Approaches for reducing the organization's energy demands, energy intensity, and energy costs
7. Ways to leverage future societal energy needs and issues for competitive advantage by creating new products and services that add value and satisfy customer demands
8. Planning for the utilization of innovation and new technologies to address energy issues

In light of the linkage between energy usage, greenhouse gas emissions, and potential climate change there is increasing public scrutiny of and concern about industry's energy practices. Because of this growing public concern, it is important for all organizations to understand their energy consumption, the environmental impacts of their energy use, and the implications of the company's energy choices. Therefore, as in other improvement activities, a key initial step in any Lean and Green energy reduction initiative is to gather the necessary energy usage and loss data. An ideal method for obtaining accurate, real-time energy usage data is the installation of wireless submeters for different equipment or areas within a process or facility. Other approaches for determining energy usage include engineering calculations and the use of utility monitoring data. However it is obtained, accurate and detailed energy data will enable the organization to develop an energy inventory (Figure 8.7) highlighting where, when, and how energy is consumed in the business. The energy use data can also be utilized in value stream mapping to provide a visual picture of the organization's energy use in relation to other process parameters.[22] This information can then be utilized to identify suitable targets for energy improvement projects or *kaizens*, and to make informed decisions about energy matters.

Integrating Lean methodologies into an organization's operating model, its safety, health, and environmental management system, and its sustainability

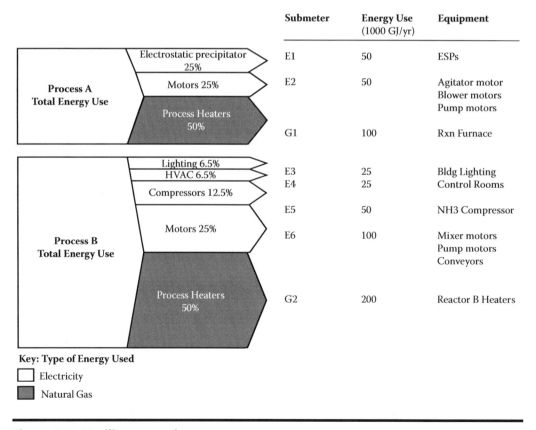

Figure 8.7   Facility energy inventory.

initiatives can enable an enterprise to become a paragon of business excellence where the "Lean, Green, and Serene" vision becomes reality. The organization is "Lean" or efficient because it adopts Lean ways of working that enable it to systematically eliminate workplace losses, and to increase operational efficiency. The business is "Green" because it employs and leverages Lean practices to conserve natural resources, reduce its environmental footprint, and operate in a sustainable fashion. The workplace is "Serene" or safe because Lean principles have been applied to safety in order to eliminate occupational accidents and incidents. Establishing safety as an organizational value firmly integrated into the company's management system makes employees feel calm, secure, and valued. In today's competitive world the synthesis of Lean, SHE, and sustainability is a recipe for enduring market success for business stakeholders, for society, and for the environment. The days of companies operating in the "doing business as usual" mode are numbered. The "Lean, Green, and Serene" approach offers a real opportunity for businesses to do well by doing good.

Endnotes

1. Gordon, Pamela J. (2001). *Lean and Green: Profit for Your Workplace and the Environment*. Berrett-Koehler, San Francisco.
2. Siegel, Yakir and Amy Longsworth (2009). Sustainability for CEOs. *Chief Executive*. (Jan/Feb): 238.
3. Hawken, Paul (1993). *The Ecology of Commerce: A Declaration of Sustainability*. HarperCollins, New York.
4. Hawken, Paul, Amory Lovins, and L. Hunter Lovins (1999). *Natural Capitalism: Creating the Next Industrial Revolution*. Little, Brown, New York.
5. McDonough, William and Michael Braungart (2002). *Cradle to Cradle: Remaking the Way We Make Things*. North Point Press, New York.
6. Anderson, Ray (2009). *Confessions of a Radical Industrialist: Profits, People, Purpose—Doing Business by Respecting the Earth*. St. Martin's Press, New York.
7. Interface Global (2009). "Interface's values are our guiding principles." Retrieved on Nov. 12, 2009 from: http://www.interfaceglobal.com/Company/Mission-Vision.aspx
8. Interface Global (2009). "Global ecometerics." Retrieved on Nov. 12, 2009 from: http://www.interfaceglobal.com/getdoc/7e96b54e-ad49-4eff-9877-38a55df0396d/Global-EcoMetrics.aspx
9. Anderson, op. cit., p. 5.
10. Elkington, J. (1994). Towards the sustainable corporation: Win-win-win business strategies for sustainable development. *California Management Review* 36, 2: 90–100.

11. Wolf, L. (2008). Triple bottom line networks (SVN, BALLE, B Corporation). ElephantJournal.com. November 25. Retrieved on October 10, 2009 from: http://www.elephantjournal.com/2008/11/triple-bottom-line-business-networks-svn-balle-b-corporation/

12. Womack, James (2009). In Peter Hines, "Lean & green." Retrieved on October 10, 2009 from: http://www.slideshare.net/PeterHines/Lean-Green-viewpoint-from-Professor-Peter-Hines

13. Hines, Peter (2009). "Lean & green." Retrieved on October 10, 2009 from: http://www.slideshare.net/PeterHines/Lean-Green-viewpoint-from-Professor-Peter-Hines

14. Lubin, David A. and Daniel C. Esty (2010). The sustainability imperative. *Harvard Business Review.* 88(5), 43–50.

15. Lacy, Peter, Tim Cooper, Rob Haywood, and Lisa Newberger (2010). "A new era of sustainability. UN Global Compact—Accenture CEO Study 2010." Retrieved from: https://microsite.accenture.com/sustainability/Documents/Accenture_UNGC_Study_2010.pdf

16. Infor (2007). "Going green: How environmentally conscious practices and products present a profitable future today." Infor. Alpharetta, Georgia. Retrieved from: www.infor.com

17. Lubin and Esty, op. cit., p. 47.

18. Apple (2010). "Apple and the environment: The story behind Apple's environmental footprint." Retrieved on September 15, 2010 from: http://www.apple.com/environment/complete-lifecycle/#product

19. Unilever (2010). "Unilever sustainable living plan: Small actions, big difference." Retrieved on November 24, 2010 from: http://www.unilever.com/images/UnileverSustainableLivingPlan_tcm13-239379.pdf

20. US EPA (2007). "The Lean environmental toolkit." United States Environmental Protection Agency Publication EPA-100-K-06-003 Revised October, 2007. Retrieved from: www.epa.gov/lean

21. GBN (2007). "Energy strategy for the road ahead: Scenario thinking for business executives and corporate boards." Global Business Network. San Francisco. Retrieved on Sept. 15, 2010 from: http://www.energystar.gov/ia/business/GBN_Energy_Strategy.pdf

22. US EPA (2007). "The Lean and Energy Toolkit." United States Environmental Protection Agency Publication EPA-100-K-07-003, Revised October, 2007. Retrieved from: www.epa.gov/lean

Bibliography

Alderton, Margo (2007). "Green is gold, according to Goldman Sachs study." CRO Corporate Responsibility Officer. June 2007. Retrieved from: http://www. thecro.com/node/490. Original paper: Ling, Anthony et al. Goldman Sachs Introducing GS Sustain. June 22, 2007 at: http://www.unglobalcompact.org/ docs/summit2007/gs_esg_embargoed_until030707pdf.pdf

Amparo, Oliver, José Manuel Tomás, and Alistair Cheyne (2006). Safety climate: Its nature and predictive power. *Psychology in Spain*. 10: 28–36.

Anderson, Ray (2009). *Confessions of a Radical Industrialist: Profits, People, Purpose—Doing Business by Respecting the Earth*. St. Martin's Press, New York.

Apple (2010). "Apple and the environment: The story behind Apple's environmental footprint." Retrieved on September 15, 2010 from: http://www.apple.com/ environment/complete-lifecycle/#product

ASME (ANSI) A13.1-2007 (2007). *Scheme for the Identification of Piping Systems*. American Society of Mechanical Engineers, New York.

Barrett, Richard (2009). "Values based leadership: Why is it important for the future of your organization?" Barrett Values Centre. Retrieved on Sept. 1, 2009 from: www.valuescentre.com/docs/ValuesBasedLeadership.pdf

Baue, William (2002). "Eco-efficient pharmaceutical companies have higher share value." *Social Funds*, July 3. http://www.socialfunds.com/news/article. cgi/873.html. Original paper: Innovest EcoValue 21 Study. The global pharmaceutical industry: Uncovering hidden value potential for strategic investors.

Chase, R. B. and D. M. Stewart (1994). Make your service fail-safe. *Sloan Management Review*. (Spring): 35(3), 35–44.

City of Atlanta (2007). Hartsfield Airport fact sheet. Atlanta. Feb. 2007.

Digenti, Dori (1996). "Zen learning: A new approach to creating multi-skilled workers." Center for International Studies, Massachusetts Institute of Technology Japan Program. Boston, MA. Retrieved on April 14, 2010 from: http://dspace. mit.edu/bitstream/handle/1721.1/7574/JP-WP-96-29.pdf?sequence=1

DOE. (1992). DOE performance indicators guidance document. DOE standard 1048–92. U.S. Department of Energy. Washington, DC.

Drucker, Peter (1972). *Technology, Management, and Society*. Harper and Row, New York.

Edwards, Andres (2005). *The Sustainability Revolution*. New Society, Gabriola Island, BC, Canada.

Elkington, J. (1994). Towards the sustainable corporation: Win-win-win business strategies for sustainable development. *California Management Review*. 36, 2: 90–100.

Flin, R., K. Mearns, P. O'Conner, and R. Bryden (2000). Measuring safety climate: Identifying the common features. *Safety Science* 34: 177–192.

FMEA-FMECA.com (2010). "Your guide to FMEA—FMECA information." Retrieved on March 7, 2010 from: http://www.fmea-fmeca.com/index.html

Ford, Henry (1922). *My Life and Work*. Nevins and Hill Publications, LaVergne, TN.

Fox, Ron (1995). "The Aikido FAQ: Shu Ha Ri." *The Iaido Newsletter*. 7, 2: 1. Retrieved on April 28, 2010 from: http://www.aikidofaq.com/essays/tin/shuhari.html

Gawande, Atul (2009). *Checklist Manifesto: How to Get Things Done Right*. Metropolitan, New York.

GBN (2007). "Energy strategy for the road ahead: Scenario thinking for business executives and corporate boards." Global Business Network. San Francisco. Retrieved on Sept. 15, 2010 from: http://www.energystar.gov/ia/business/GBN_Energy_Strategy.pdf

GEMI—Global Environmental Management Initiative (February 2004). "Clear advantage: building shareholder value, environment—Value to the investor." Retrieved from: http://www.gemi.org/resources/GEMI%20Clear%20Advantage.pdfhttp://www.gemi.org/supplychain/C2.htm

Goldman Sachs JBWere (October, 2007). "Media release. Goldman Sachs finds valuation links in workplace health and safety data." Retrieved from: http://www.gsjbw.com/documents/About/MediaRoom/GSJBW-WHS-Report-Media-Release.pdf

Gordon, Pamela J. (2001). *Lean and Green: Profit for Your Workplace and the Environment*. Berrett-Koehler, San Francisco.

Graphic Products, Inc. (2009) "The 5S philosophy." Beaverton, OR. Retrieved Sept. 1, 2009 from: www.GraphicProducts.com

Greif, Michel (1991). *The Visual Factory*. Productivity Press, Portland, OR.

Hallowell, Matthew, Anthony Veltri, and Stephen Johnson (November, 2009). Safety & Lean: One manufacturer's lessons learned and best practices. *Professional Safety*. 54, 11 (Nov.): 22.

Harvard Business Review (October, 2007). "Climate business, business climate." Forethought Special Report. pp. 1–17. Retrieved from: http://www.erb.umich.edu/News-and-Events/news-events-pics/HBR-Oct07.pdf

Hawken, Paul (1993). *The Ecology of Commerce: A Declaration of Sustainability*. HarperCollins, New York.

Hawken, Paul, Amory Lovins, and L. Hunter Lovins (1999). *Natural Capitalism: Creating the Next Industrial Revolution*. Little, Brown, New York.

Hess, Edward D. and Kim S. Cameron (2006). *Leading with Values: Positivity, Virtue and High Performance*. Cambridge University Press, Cambridge, UK.

Hines, Peter (2009). "Lean & green." Retrieved on October 10, 2009 from: http://www.slideshare.net/PeterHines/Lean-Green-viewpoint-from-Professor-Peter-Hines

Hutchens, S. (2010). "Using ISO 9001 or ISO 14001 to gain a competitive advantage." Intertek white paper. September. Retrieved from: http://www.intertek.com/WorkArea/DownloadAsset.aspx?id=4431

Imai, Masaaki (1997). *Gemba Kaizen: A Commonsense, Low-Cost Approach to Management.* McGraw-Hill, New York, p. 58.

Infor (2007a). "Going green: Practice for a profitable future." Infor. Alpharetta, GA. Retrieved from: http://www.infor.com/goinggreen/

Infor (2007b). "Going green: How environmentally conscious practices and products present a profitable future today." Infor. Alpharetta, Georgia. Retrieved from: www.infor.com

Interface Global (2009a). "Global ecometerics." Retrieved on Nov. 12, 2009 from: http://www.interfaceglobal.com/getdoc/7e96b54e-ad49-4eff-9877-38a55df0396d/Global-EcoMetrics.aspx

Interface Global. (2009b). "Interface's values are our guiding principles." Retrieved on Nov. 12, 2009 from: http://www.interfaceglobal.com/Company/Mission-Vision.aspx

ISO (2004). ISO 14001:2004 Environmental management systems—Requirements with guidance for use. International Organization for Standardization. Geneva.

Johnson, Dave (2003). Perception is reality. *Industrial Safety and Hygiene News ISHN E-NEWS.* Friday, October 3, 2, p. 30.

Johnson, Steven (2006). *The Ghost Map: The Story of London's Most Terrifying Epidemic—And How It Changed Science, Cities and the Modern World.* Riverhead Books/Penguin, New York.

Kaplan, Robert and David Norton (1992). The balanced scorecard—measures that drive performance. *Harvard Business Review.* (Jan–Feb): 80(1), 71–80.

Kaplan, Robert and David Norton (1993). Putting the balanced scorecard to work. *Harvard Business Review.* (Sep–Oct): 71(5), 134–147.

Kaplan, Robert and David Norton (December 23, 2002). "Partnering and the balanced scorecard." *Harvard Business Review.* Retrieved from: http://hbswk.hbs.edu/item/3231.html

Klentz, Trevor (1999). *HAZOP and HAZAN.* Institution of Chemical Engineers. Rugby, UK.

Lacy, Peter, Tim Cooper, Rob Haywood, and Lisa Newberger (2010). "A new era of sustainability. UN Global Compact—Accenture CEO Study 2010." Retrieved from: https://microsite.accenture.com/sustainability/Documents/Accenture_UNGC_Study_2010.pdf

Leflar, James A. (2001). *Practical TPM: Successful Equipment Management at Agilent Technologies.* Productivity Press, Portland, OR.

Liberty Mutual (2001). "A majority of U.S. businesses report workplace safety delivers a return on investment." Liberty Mutual News Release. Boston. August 28. Retrieved from: www.libertymutual.com

Liker, Jeffrey (2004) *The Toyota Way. 14 Management Principles from the World's Greatest Manufacturer.* McGraw-Hill, New York.

Low, Jonathan, Pam Cohen, and Pamela Kalafut (2002). *Invisible Advantage: How Intangibles Are Driving Business Performance.* Persius Press, Basic, Cambridge, MA.

Lu, David (1986). *Kanban Just-in-Time at Toyota: Management Begins at the Workplace.* Productivity Press, Portland, OR.

Lubin, David A. and Daniel C. Esty (2010). The sustainability imperative. *Harvard Business Review.* 88, 5 (May): 42–49.

Majer, Kenneth (2004). *Values-Based Leadership: A Revolutionary Approach to Business Success and Personal Prosperity.* Majer Communications, San Diego.

Manuele, Fred A. (2003). *On the Practice of Safety.* John Wiley & Sons, Hoboken, NJ.

Manuele, Fred A. (2007). Lean concepts: Opportunities for safety professionals. *Professional Safety.* 52, 8 (Aug): 28–34.

Martin, Raymond (1998). "ISO 14001" guidance manual. National Center for Environmental Decision Making Research. Technical Report NCEDR/98-06. Retrieved on January 20, 2010 from: http://www.usistf.org/download/ISMS_Downloads/ISO14001.pdf

Maslow, Abraham H. (1943). A theory of human motivation. *Psychological Review.* 50: 370–396.

McDermott, Robin, R. Mikulak, and M. Beauregard 1996. *The Basics of FMEA.* Productivity Press, Portland, OR.

McDonough, William and Michael Braungart (2002). *Cradle to Cradle: Remaking the Way We Make Things.* North Point Press, New York.

Millard, Lorraine (1999). *5S for Safety: New Eyes for the Shop Floor.* Primedia Workplace Learning, New York.

Nagamachi, Mitsuo (2008). Shiromiso—Akamiso accident prevention management based on brain theory. Hiroshima International University. *2008 International Conference on Productivity & Quality Research.*

Nakajima, Seiichi (1989). *TPM Development Program: Implementing Total Productive Maintenance.* Productivity Press, Cambridge, MA.

Nakano, Kinjiro (2003). *Planned Maintenance Keikaku Hozen: Comprehensive Approach to Zero Breakdowns.* Japan Institute of Plant Maintenance. Tokyo, Japan.

National Center for Health Statistics (2006). *Fast stats A to Z.* CDC. Atlanta. December.

National Safety Council (2005). *Injury Facts 2004 Edition.* NSC Press, Chicago.

NFPA 704-2007. (2007). *Standard System for the Identification of the Hazards of Materials for Emergency Response.* National Fire Protection Association. Quincy, MA.

OHSAS (2007). OHSAS 18001-2007: "Occupational health and safety management, requirements." British Standards Institution (BSI). Retrieved from: http://shop.bsigroup.com/ProductDetail/?pid=000000000030148086

Owen, Mal (1989). *SPC and Continuous Improvement.* IFS, Kempston, UK.

Petersen, Dan (1988). *Safety Management: A Human Approach*. Aloray, Goshen, NY.

Peterson, Ivars (1996). *Fatal Defect*. Random House, New York, p. 111.

Plato (427–347 BC) *The Republic*. Penguin Classics, New York, September, 1955.

Prevette, S. (2006). Charting safety performance: Combining statistical tools provides quality data. *Professional Safety*. 51, 5 (May): 34–41.

Prevette, S., W. Previty, A. Umek, and C. Hayes (2009). Implementing performance trending at two Department of Energy sites. *WM Conference*, March, 2009.

Productivity Europe (1998). *The 5S Improvement Handbook*. Productivity Europe, Bedford, UK.

Robinson, Alan (1991). *Continuous Improvement in Operations*. Productivity Press, Portland, OR.

Robinson, Charles and Andrew Ginder (1995). *Implementing TPM—The North American Experience*. Productivity Press, Portland, OR.

Royal Society of Chemistry (2010). "Environment, health and safety committee note on: Hazards and operability studies (HAZOP)." Retrieved on February 5, 2010 from: http://www.rsc.org/images/HAZOPs_V2_190707_tcm18-95646.pdf

Rule Book of the New York and Erie Railroad (1854). "Great Aviation Quotes." Retrieved on Sept 29, 2010 from: http://www.skygod.com/quotes/safety

Savitz, Andrew and Karl Weber (2006). *The Triple Bottom Line*. John Wiley & Sons, San Francisco.

Schweitzer, Melodie (2007). Creating a safety culture through felt leadership. *Industrial Hygiene News*. November. Retrieved from: http://www.rimbach.com/scripts/Article/IHN/Number.idc?Number=113

Senge, Peter (2010). *The Necessary Revolution: Working Together to Create a Sustainable World*. Broadway, New York.

Shirosi, Kunio (1992). *TPM for Workshop Leaders*. Productivity Press, Portland, OR.

Shirosi, Kunio (1996). *Total Productive Maintenance: New Implementation Program in Fabrication and Assembly Industries*. Japan Institute of Plant Maintenance (JIPM). Tokyo, Japan.

Siegel, Yakir and Amy Longsworth (2009). Sustainability for CEOs. *Chief Executive*. (Jan/Feb): 238.

Smith, Rick L. (2004). "All breakdowns can be prevented." *Fabricating and Metalworking Magazine*. Retrieved on September 21, 2009 from: http://www.technicalchange.com/pdfs/why-are-these-men-smiling.pdf

Soin, Sarv Singh (1998). *Total Quality Essential: Using Quality Tools and Systems to Improve and Manage Your Business*. McGraw-Hill Professional, New York.

Stewart, James M. (1993). Future state visioning—A powerful leadership process. *Long Range Planning*. 26, 6: 89–98.

Stewart, James M. (2002). *Managing for World Class Safety*. Wiley-Interscience, New York

Sullivan, Louis H. (1896). "The tall office building artistically considered." *Lippincott's Magazine*, March 1896 in MIT Open Courseware. Retrieved from: http://ocw.mit.edu/OcwWeb/Civil-and-Environmental-Engineering/1-012Spring2002/Readings/detail/-The-Tall-Office-Building-Artistically-Considered-.htm

Sun Tzu (2005). *The Art of War*. English translation by Lionel Giles. El Paso Norte Press, El Paso, TX.

Suzuki, Tokutaro (1994). *TPM in Process Industries*. Productivity Press, New York.

Thomas, Jim and John Harris (February 2004). "Clear advantage: Building shareholder value. Environment: Value to the investor." GEMI (Global Environmental Management Initiative). Retrieved from: http://www.gemi.org/resources/GEMI%20Clear%20Advantage.pdf

Unilever (2010). "Unilever sustainable living plan: Small actions, big difference." Retrieved on November 24, 2010 from: http://www.unilever.com/images/UnileverSustainableLivingPlan_tcm13-239379.pdf

U.S. EPA (2007a). "The Lean and energy toolkit." United States Environmental Protection Agency Publication EPA-100-K-07-003, Revised October, 2007. Retrieved from: www.epa.gov/lean

U.S. EPA (2007b). "The Lean environmental toolkit." United States Environmental Protection Agency Publication EPA-100-K-06-003 Revised October, 2007. Retrieved from: www.epa.gov/lean

Ventana Research (2007). Strategies to run a Lean supply chain. How principles of Lean manufacturing transfer benefits to operations. Ventana Research White Paper. San Mateo, CA.

Vogel, David (2005). *The Market for Virtue: The Potential and Limits of Corporate Social Responsibility*. The Brookings Institution. Washington, DC.

Wattleton, Faye (2009). WhatQuote.com. Retrieved Sept. 1, 2009 from: http://www.whatquote.com/quotes/Faye-Wattleton/16448-The-only-safe-ship-i.htm

White, Ander and Matthew Kiernan (September 2004). "Innovest Strategic Value Advisors. Corporate environmental governance: A study into the influence of environmental governance and financial performance." Environmental Agency Bristol, UK. Retrieved from: http://publications.environment-agency.gov.uk/pdf/GEHO0904BKFE-e-e.pdf

Whitman, Mark (Nov. 2005). "The culture of safety: No one gets hurt today." *The Police Chief*. 72, 11. Retrieved from: http://policechiefmagazine.org/magazine/index.cfm?fuseaction=display_arch&article_id=737&issue_id=112005

Wiegmann, Douglas A., H. Zhang, T. von Thaden, and A. Mitchell (2002). "A synthesis of safety culture and safety climate research." University of Illinois at Urbana-Champaign and Federal Aviation Administration Technical Report ARL-02-3/FAA-02-2. June. Retrieved from: http://www.humanfactors.uiuc.edu/Reports&PapersPDFs/TechReport/02-03.pdf

Willard, Bob (2005). *The Next Sustainability Wave; Building Boardroom Buy-In*. New Society, Gabriola Island, BC, Canada.

Winter, Jessica (2007). "A world without waste." *Boston Globe*. March 11, 2007. Retrieved on September 15, 2010 from: http://www.boston.com/news/education/higher/articles/2007/03/11/a_world_without_waste/

Wolf, L. (2008). "Triple bottom line networks (SVN, BALLE, B Corporation)." ElephantJournal.com. November 25. Retrieved on October 10, 2009 from: http://www.elephantjournal.com/2008/11/triple-bottom-line-business-networks-svn-balle-b-corporation/

Womack, James (2009). In Peter Hines, "Lean & green." Retrieved on October 10, 2009 from: http://www.slideshare.net/PeterHines/Lean-Green-viewpoint-from-Professor-Peter-Hines

Womack, James and Daniel Jones (1996). *Lean Thinking: Banish Waste and Create Wealth in Your Corporation. Part 1: Lean Principles.* Simon & Schuster, New York.

Womack, James and Daniel Jones (2005). *Lean Solutions: How Companies and Customers Can Create Value and Wealth Together.* Free Press, a division of Simon & Schuster. New York.

Womack, James et al. (1990). *The Machine That Changed the World: The Story of Lean Production.* Rawson Associates, a division of Macmillan, New York.

Index

39714443R00125

Made in the USA
Middletown, DE
23 January 2017